The Handbook of Information Security for Advanced Neuroprosthetics

Second Edition

Matthew E. Gladden

SYNTHYPNION
academic

The Handbook of Information Security for Advanced Neuroprosthetics
(Second Edition)

Published in the United States of America
by Synthypnion Academic, an imprint of Synthypnion Press LLC

Synthypnion Press LLC
Indianapolis, IN 46227
http://www.synthypnionpress.com

SYNTHYPNION academic
synthypnionpress.com

SYNTHYPNION
academic

ISBN 978-1-944373-09-2 (hardcover print edition)
ISBN 978-1-944373-14-6 (paperback print edition)
ISBN 978-1-944373-10-8 (ebook)
10 9 8 7 6 5 4 3 2 1
February 2017

To all those who make use of these technologies,
whether out of necessity or by choice –
that such tools might be sources of health,
freedom, and full human development.

Brief Table of Contents

Detailed Table of Contents

Preface

This second edition of *The Handbook of Information Security for Advanced Neuroprosthetics* updates the previous edition in a number of significant ways. To begin with, two texts that appeared in the first edition (Chapter 4 on "An Information Security Device Ontology for Advanced Neuroprosthetics" and the appendix titled "Biocybernetic Classification of Advanced Neuroprosthetic Devices") have been omitted from this volume; readers who are interested in such texts may wish to consult Chapters 1-3 of the newly published book *Neuroprosthetic Supersystems Architecture: Considerations for the Design and Management of Neurocybernetically Augmented Organizations*, which consider the same material in greater depth. (And, indeed, the entire volume of *Neuroprosthetic Supersystems Architecture* may be understood as providing a foundation and introduction to the present book, as it presents a general overview of the relationship of posthumanizing neuroprostheses to organizational structures, processes, and systems – a topic which this volume then investigates through the more specific lens of information security.)

Second, a new appendix has been added to this book in the form of "Information Security Concerns as a Catalyst for the Development of Implantable Cognitive Neuroprostheses," which was originally published in the 9th *Annual EuroMed Academy of Business (EMAB) Conference: Innovation, Entrepreneurship and Digital Ecosystems (EUROMED 2016) Book of Proceedings*, edited by Demetris Vrontis, Yaakov Weber, and Evangelos Tsoukatos, pp. 891-904; Engomi: EuroMed Press, 2016.

Finally, the first edition of this book dedicated Chapters 7-9 to the consideration of *management*, *operational*, and *technical* controls for advanced neuroprosthetic systems. This second edition considers much the same material but organizes it in a different way, grouping such measures into *preventive*, *detective*, and *corrective* controls. In this we follow the lead of publications such as *NIST Special Publication 800-53, Revision 4: Security and Privacy Controls for Federal Information Systems and Organizations* (2013), which has removed from its catalog of security controls the explicit categorization of such measures as management, operational, or technical controls, due to the fact that so many controls reflect aspects of more than one category and their forced classification in any single category would be rather arbitrary. It is hoped that the new classification of controls as preventive, detective, or corrective and compensating provides a more rational and useful way of conceptualizing such measures.

Throughout the process of researching and writing both editions of this book, I have benefitted from the input and support of many individuals who contributed in one way or another to the successful completion of the project. Among that group are many scholars who have shared questions, feedback, and suggestions at the conferences at which I presented material that was eventually incorporated into this text. For the insights that they have shared I would especially like to thank Bartosz Kłoda-Staniecko, Magdalena Szczepocka, Michał Kłosiński, Agata Kowalewska, Piotr Toczyski, Alan N. Shapiro, Krzysztof Maj, Katarzyna Marak, Miłosz Markocki, Jakub Krogulec, Dawid Junke, and Sven Dwulecki. I am also grateful to all the faculty and administrators of the University of Warsaw's Digital Economy Lab and the Institute of Computer Science of the Polish Academy of Sciences – particularly Serge Pukas and Paulina Krystosiak, who provided me with much encouragement in my research, and Robert Pająk, who offered a number of valuable insights and questions during my research.

I am also grateful to the faculty and staff of Georgetown University's School of Continuing Studies, the current and former faculty and staff of the University's Department of Psychology, and all of the research fellows and staff with whom I worked at the Woodstock Theological Center. I am especially indebted to Father John Haughey, S.J., for his insights relating to various aspects of transhumanism; to Terry Armstrong, for his generous encouragement, his knowledge of artificial intelligence and organizational management, and his good example; and to Father Gap Lo Biondo, S.J., for more than the printed page can contain. I am also thankful to friends and colleagues such as AJ Johnson and Nathan Fouts (with whom over the years I have enjoyed helpful conversations about topics of cybernetics, posthumanism, and information security whose fruits have found a home in this text) and Tom Rijntjes (for his inspiring example of ingenuity, intellectual entrepreneurship, and tireless labor). I owe a boundless debt to my parents, my brother, and my wife for their unceasing encouragement and support. And finally, I thank all of those whom I have forgotten to mention by name not because their contribution was so slight but because it was so great that I have come to take it (and them) for granted.

Together, all of the individuals mentioned above have made it possible for me to prepare this book; they have contributed immensely to whatever value this text might hold for those concerned about information security for advanced neuroprosthetic devices and systems. Whatever flaws, biases, and limitations the book may possess are the fault not of any of the persons mentioned above but are my responsibility alone.

Matthew E. Gladden
Pruszków, February 16, 2017

Introduction

The Purpose and Organization of This Text

How can one provide adequate information security for an implantable device that is integrated into the neural processes of a human brain – and for the larger biocybernetic system that such a device creates with its human host? That is the essential question that has given rise to this book, which is the first text dedicated to studying the issue comprehensively from both theoretical and practical perspectives.

The material presented in this volume is organized into two main parts plus an appendix. The first part of this book provides an introduction to key themes and questions that provide the foundation for the entire text and which will recur throughout the volume in many different contexts. In **Chapter One**, we present an introduction to neuroprosthetic devices and systems that explores both the state of the art of sensory, motor, and cognitive neuroprostheses that are currently in use as well as more sophisticated kinds of neuroprosthetic technologies that are being actively pursued or that are expected to be developed in the future. This overview takes us from the contemporary world of neuroprosthetic devices that have been designed primarily for purposes of therapeutic treatment of medical disorders and the restoration of natural human abilities lost due to illness or injury to an emerging future world in which neuroprosthetic devices offer the possibility of augmenting and transforming the capacities of their users in such a way that they can perhaps best be described as 'posthumanizing' technologies.

In **Chapter Two**, we present an introduction to basic principles and practices within the complex, diverse, and dynamic field of information security. Concepts such as the CIA Triad; the nature of administrative, logical, and physical security controls; the role of access controls in performing user authentication and authorization; and the differences between vulnerabilities, threats, and risks are all discussed. We also highlight the different forms that information security can take when pursued by a large organization as opposed to, say, the individual user of a consumer electronics device.

In **Chapter Three**, we consider unique challenges, problems, and opportunities that arise when one attempts to apply the general principles and practices of information security to the somewhat idiosyncratic domain of advanced neuroprosthetics. Issues discussed include the distinction between a device and its host-device system; the need for a device to provide free access to outside parties during medical emergencies while rigorously restricting access to outside parties at other times; challenges that arise relating to implanted devices' limited power supply and processing and storage capacities, physical inaccessibility, and reliance on wireless communication; new kinds of biometrics that can be utilized by neuroprosthetic devices; complications and opportunities arising from the use of nontraditional computing structures and platforms such as biomolecular computing and nanorobotic swarms; and psychological, social, and ethical concerns that arise relating to the agency, autonomy, and personal identity of human beings possessing advanced neuroprostheses. Having considered such issues, we discuss why traditional concepts of information security that are often applied to general-purpose computing and information systems are inadequate to address the realities of advanced neuroprosthetic host-device systems and why the creation of new specialized conceptual models of information security for advanced neuroprosthetics is urgently required.

In **Chapter Four** we develop a two-dimensional cognitional security framework for advanced neuroprosthetic devices that takes into account not only the information security needs of a neuroprosthesis itself but also those of the host-device system that the device creates through its integration into the neural circuitry of its human host. The framework first describes nine information security goals or attributes – namely, confidentiality, integrity, availability, possession, authenticity, utility, distinguishability, rejectability, and autonomy. The framework considers how the pursuit of these security goals for a host-device system can be advanced (or subverted) at three different levels, in which the human host of a neuroprosthetic device is considered in his or her role as: 1) a sapient metavolitional agent; 2) an embodied embedded organism; and 3) a social and economic actor. This framework shares some common elements with classical models of information security goals that were formulated for general-purpose computing and information systems, but it also proposes new elements to address the unique nature of advanced neuroprostheses.

The second part of the book discusses practical aspects of developing and implementing information system security plans for advanced neuroprosthetic devices, either within the context of a large organization or for an individual consumer who is utilizing such a device. In **Chapter Five** we describe how responsibilities for planning and implementing information security practices and mechanisms are typically allocated among individuals filling different roles. We then note the unique forms that these roles and responsibilities can take when the focus of their activities is ensuring information security for advanced neuroprostheses. The next three chapters explore how security controls relate to advanced neuroprosthetic devices and systems by considering in detail the controls described in texts like *NIST Special Publication 800-53, Rev. 4: Security and Privacy Controls for Federal Information Systems and Organizations*, produced by the National Institute of Standards & Technology in 2013. We address most of the security controls noted in *NIST SP 800-53*, exploring how particular controls that were designed for use with general-purpose computing and information systems may either: 1) become *more critical* for ensuring information security when the information system to which the control is being applied is an advanced neuroprosthetic device; 2) become *less important* or potentially even irrelevant and inapplicable for an advanced neuroprosthetic device; or 3) may take on *new and radically* different forms when applied to advanced neuroprosthetic devices. **Chapter Six** focuses on preventive controls, **Chapter Seven** considers detective controls, and **Chapter Eight** addresses corrective or compensating controls.

Finally, in the **Appendix** we consider the ways in which InfoSec concerns may serve as a catalyst for the development of implantable cognitive neuroprostheses (ICNs). In the case of ICNs that are integrated with the neural circuitry of their human hosts, there is a widespread presumption that InfoSec concerns serve only as limiting factors that can complicate, impede, or preclude the development and deployment of such devices. However, we argue that when appropriately conceptualized, InfoSec concerns may also serve as drivers that can spur the creation and adoption of such technologies. A framework is formulated that describes seven types of actors whose participation is required in order for ICNs to be adopted; namely, their 1) producers, 2) regulators, 3) funders, 4) installers, 5) human hosts, 6) operators, and 7) maintainers. By mapping onto this framework InfoSec issues raised in industry standards and other literature, it is shown that for each actor in the process, concerns about information security can either disincentivize or incentivize the actor to advance the development and deployment of ICNs for purposes of therapy or human enhancement. For example, it is shown that ICNs can strengthen the integrity, availability, and utility of information stored in the memories of persons suffering from certain neurological conditions and may enhance information security for society as a whole by providing new tools for military, law enforcement, medical, or corporate personnel who provide critical InfoSec services.

As the first book dedicated to studying these issues comprehensively from both theoretical and practical perspectives, it is hoped that this volume can serve as a resource for specialized studies in information security, cybernetics, bioethics, and biotechnology; as reference for researchers and practitioners working to ensure information security for advanced neuroprostheses; and as a starting point for all those who are seeking to explore the subject and are in search of practical and conceptual frameworks that can guide the development of this emerging field.

Part I

Background and Foundations

Chapter One

An Introduction to Advanced Neuroprosthetics

Abstract. This text presents an introduction to neuroprosthetic devices and systems that explores both the state of the art of sensory, motor, and cognitive neuroprostheses that are currently in use as well as more sophisticated kinds of neuroprosthetic technologies that are being actively pursued or that are expected to be developed in the future. This overview takes us from the contemporary world of neuroprostheses that have been designed primarily for purposes of therapeutic treatment of medical disorders and the restoration of natural human abilities lost due to illness or injury to an emerging future world in which neuroprosthetic devices offer the possibility of augmenting and transforming the capacities of their users in such a way that they can perhaps best be described as 'posthumanizing' technologies.

I. Overview of current neuroprosthetic devices

The integration of human beings with computers at both the physical and cognitive levels is growing ever deeper, as new technologies are developed and the daily routines of our human existence adapt to incorporate these new means of experiencing and shaping reality. Traditionally, **human-computer interaction (HCI)** has relied on tools that are external to the human body, such as keyboards, mice, computer screens, and speakers. In recent years, the emergence of mobile and wearable technologies such as smartphones, smartwatches, and virtual reality headsets has created a new range of devices that are more intimately connected with the bodies of their human users. But for a growing population of persons, computerized information systems are no longer technologies that simply exist outside of – or even on the surface of – their bodies; for these persons, computing technologies have passed through the boundaries of the human body and have come to exist and to operate within their physical being. For example, an increasing number of human beings now house within their bodies implantable computers that are active and functioning continually as those persons go about their everyday activities. Such implantable computers often form key components of implantable medical devices (IMDs) such as defibrillators, pacemakers, deep brain stimulators, retinal and cochlear implants, or diagnostic devices such as body area

networks (BANs) or body sensor networks (BSNs). Some of the more sophisticated forms of implantable RFID transponders also function as implantable
computers.[1] Such implantable computers are increasingly serving as sites for
the reception, generation, storage, processing, and transmission of large
quantities of highly sensitive information[2] regarding almost every aspect of
the lives of their human hosts, including their hosts' everyday interactions
with the environment (including interactions with other human beings),
their internal biological processes, and even their cognitive activity.

One kind of computer that becomes linked with a particular human being's organism in an especially powerful and intimate way is a **neuroprosthetic
device** that is integrated directly into the body's neural circuitry.[3] A neuroprosthetic device may either be physically inserted into the brain, as in the
case of many kinds of brain implants already in use, or it could potentially
surround the brain, as in the case of a full cyborg body of the sort envisioned
by some researchers and futurologists.[4] Neuroprosthetic devices increasingly
operate in rich and complex biocybernetic and neurocybernetic control loops
with the body and mind of their human host, allowing the host's cognitive
activity to be detected, analyzed, and interpreted for use in exercising real-
time control over computers or robotic devices.[5]

The terminology used to describe such devices is still quite fluid and not
always precise, as it is evolving rapidly alongside the underlying technologies.

---

[1] See Gasson et al., "Human ICT Implants: From Invasive to Pervasive" (2012), and Gasson, "ICT
Implants" (2008).

[2] See Kosta & Bowman, "Implanting Implications: Data Protection Challenges Arising from the
Use of Human ICT Implants" (2012); Li et al., "Advances and Challenges in Body Area Network"
(2011); and Rotter & Gasson, "Implantable Medical Devices: Privacy and Security Concerns"
(2012).

[3] For a discussion of circuit models as they apply to neural information processing, see Ma et al.,
"Circuit Models for Neural Information Processing" (2005). For the challenges involved with designing electrodes and other implantable electronic devices or structures that can create a sus-
tainable interface with individual neurons, see Passeraub & Thakor, "Interfacing Neural Tissue
with Microsystems" (2005). For a discussion of different technologies used to interface electronic
systems with peripheral nerves (e.g., cuff, book, or helix electrodes) or cortical neurons (e.g.,
needle arrays), see Koch, "Neural Prostheses and Biomedical Microsystems in Neurological Rehabilitation" (2007). Emerging technologies such as optogenetics used to modulate neuronal firing may make it possible to solve (or avoid) some problems relating to biocompatibility and the
degradation of tissues and electrodes experienced with conventional implanted electrode systems; see Humphreys et al., "Long Term Modulation and Control of Neuronal Firing in Excitable
Tissue Using Optogenetics" (2011).

[4] See Lebedev, "Brain-Machine Interfaces: An Overview" (2014), p. 99.

[5] See Fairclough, "Physiological Computing: Interfacing with the Human Nervous System" (2010),
and Park et al., "The Future of Neural Interface Technology" (2009).

Lebedev notes that while particular terms may be more appropriate in specific circumstances, neuroprosthetic devices and systems are often described
interchangeably as "brain-machine interfaces" (or BMIs), "neural prostheses,
brain-computer interfaces (BCIs), neural interfaces, mind-machine interfaces
and brain implants."[6] The design and use of such devices is sometimes understood as a subfield of operative neuromodulation, which involves "altering
electrically or chemically the signal transmission in the nervous system by
implanted devices in order to excite, inhibit or tune the activities of neurons
or neural networks" – something that is done typically (at least, at present) to
produce therapeutic effects.[7] Drawing on definitions offered by Lebedev[8] and
others, for the purposes of this text we can define a neuroprosthetic device as
a technological device that is integrated into the neural circuitry of a human being.
Such a definition is intentionally broad; at the same time, it is specific enough
to exclude some kinds of devices that might be considered 'neuroprosthetic
devices' by other authors writing in different contexts. We can note some key
implications of our definition as it will be employed in this text:

- A neuroprosthetic device does not need to be physically implanted
  within the body of a human host; in principle, it could function outside of its host's body (e.g., as a wearable device).

- The neuroprosthetic device must, however, be "integrated into" the
  neural circuitry of its human host. This requirement for integration
  entails a relatively rich and stable systematic connection between the
  device and some neurons within the host's body. An fMRI machine,
  for example, would thus typically not qualify as a 'neuroprosthetic
  device,' because despite the large amount of information that it generates regarding the neural activity of its host – and the effect of its
  magnetic field on the brain – it is not "integrated into" the host's neural circuitry.

- In order for a neuroprosthetic device to be integrated into the neural
  circuitry of its human host, it is not sufficient for the device to physically adjoin particular neurons or even to be completely surrounded
  by the host's neurons; rather there must be some functional interaction between the device and neurons in the host's body. Such interaction does not need to be bidirectional: a retinal prosthesis, for example, might generate and transmit an electrochemical stimulus that

---

[6] See Lebedev (2014), p. 99.

[7] See Sakas et al., "An Introduction to Neural Networks Surgery, a Field of Neuromodulation
Which Is Based on Advances in Neural Networks Science and Digitised Brain Imaging" (2007).

[8] See Lebedev (2014) and Gladden, "Enterprise Architecture for Neurocybernetically Augmented
Organizational Systems: The Impact of Posthuman Neuroprosthetics on the Creation of Strategic, Structural, Functional, Technological, and Sociocultural Alignment" (2016).

affects adjacent neurons while not being able to receive any stimulus from those neurons in return.

- A neuroprosthetic device does not need to be connected to neurons in its host's *brain*. While some existing kinds of neuroprosthetic devices indeed possess a physical interface with interneurons found in the gray matter of the human brain, a neuroprosthetic device might instead be connected to sensory or motor neurons located in limbs, sensory organs, or other parts of the body.

- A neuroprosthetic device does not need to be electronic in nature. Ongoing developments in fields such as genetic engineering, synthetic biology, bionanotechnology, and biomolecular computing are opening the door to the creation of neuroprosthetic devices that are partially or wholly composed of biological material (perhaps based on the DNA of the device's host) or other components.[9] It must, however, be a 'device' that has been developed through the use of some specific technology; in the absence of specific augmentations or modifications, a limb or organ that has simply been transplanted from another human being into its new human host would generally not qualify as a neuroprosthetic device.

A. Neuroprosthetic devices categorized by function

Existing kinds of neuroprosthetic devices have been categorized in different ways.[10] For example, a neuroprosthetic device can be classified based on the nature of its interface with the brain's neural circuitry as either **sensory, motor, bidirectional sensorimotor**, or **cognitive**.[11] We can consider each of these types of devices in turn.

1. Sensory neuroprostheses

A sensory neuroprosthesis is a neuroprosthetic device whose function is to present sense data to the mind of the device's human host.[12] Typical kinds

---

[9] For a hybrid biological-electronic interface device (or 'cultured probe') that includes a network of cultured neurons on a planar substrate, see Rutten et al., "Neural Networks on Chemically Patterned Electrode Arrays: Towards a Cultured Probe" (2007). As Rutten et al. note, such a cultured neural network would not only serve as a link between the interface's electronic components and natural neurons within the host's body but could potentially carry out its own specialized information-processing functions. Hybrid biological-electronic interface devices are also discussed in Stieglitz, "Restoration of Neurological Functions by Neuroprosthetic Technologies: Future Prospects and Trends towards Micro-, Nano-, and Biohybrid Systems" (2007).

[10] See Gladden, "Neural Implants as Gateways to Digital-Physical Ecosystems and Posthuman Socioeconomic Interaction" (2016).

[11] See Lebedev (2014).

[12] See Lebedev (2014) and Troyk & Cogan, "Sensory Neural Prostheses" (2005).

of sensory neuroprostheses already in use include cochlear implants, auditory brainstem implants,[13] and retinal prostheses.[14]

Sensory neuroprostheses may participate in different stages of a human mind's process of acquiring and perceiving sensory information. Some sensory neuroprostheses perform, participate in, or support the acquisition of raw sense data from distal stimuli. For example, a retinal prosthesis that registers the arrival of photons from the external environment and then electrically stimulates its host's natural biological retinal ganglion cells is filling such a role.

Other sensory neuroprostheses may perform the function of transmitting, translating, or transducing electrochemical signals bearing sensory information that are already present within their host's body. For example, if the retina of a human subject is still intact but part of the attached optic nerve has been damaged, a sensory neuroprosthesis could replace a portion of the optic nerve in performing the task of carrying signals from the retina to the brain. Alternatively, a sensory neuroprosthesis could be used to translate sense data from one sensory modality to another:[15] for example, auditory sense data received by hair cells in the inner ear or by a cochlear implant could be translated by the neuroprosthetic device into signals that are supplied to the optic nerve, thereby causing the incoming sounds not to be 'heard' by its host through the sensory modality of hearing but instead to be 'seen' through the sensory modality of vision – with the sounds perhaps appearing as patterns of light within a small portion of the host's field of vision, thereby creating a form of visual augmented reality.

Yet other kinds of sensory neuroprostheses might directly stimulate portions of the brain to create a particular sensory experience. For example, in the case of a human being whose optic nerve is destroyed or absent, a neuroprosthetic implant that is interconnected with neurons of its host's lateral geniculate nucleus or visual cortex could – by directly stimulating those areas – potentially cause the hostmind to experience visual phenomena that were

---

[13] Regarding cochlear implants and auditory brainstem implants, see Dormer, "Implantable electronic otologic devices for hearing rehabilitation" (2003); Cervera-Paz et al., "Auditory Brainstem Implants: Past, Present and Future Prospects" (2007); Bostrom & Sandberg, "Cognitive Enhancement: Methods, Ethics, Regulatory Challenges" (2009), p. 321; Gasson et al. (2012); Hochmair, "Cochlear Implants: Facts" (2013), and Ochsner et al., "Human, non-human, and beyond: cochlear implants in socio-technological environments" (2015).

[14] For retinal prostheses, see Weiland et al., "Retinal Prosthesis" (2005); Linsenmeier, "Retinal Bioengineering" (2005); and Viola & Patrinos, "A Neuroprosthesis for Restoring Sight" (2007).

[15] This possibility was foreseen by cyberneticists as early as the 1940s. See Wiener, *Cybernetics: Or Control and Communication in the Animal and the Machine* (1961), loc. 2784ff, and Lebedev (2014), p. 106.

not caused by any stimuli or signals present in the retina or optic nerve and which may not correspond to any distal stimuli existing in the external environment.[16] Data transmitted wirelessly to such an implant from an external computer could allow the host to experience either sense data corresponding to the 'real' environment existing outside the host's body or corresponding to some entirely 'virtual' environment whose characteristics are created and maintained by software within the computer. In all of these cases, a common trait is the fact that sensory neuroprosthetic devices are helping to present sense data to the mind of the devices' human host.

2. Motor neuroprostheses

Motor neuroprostheses, conversely, are devices that convey motor instructions – typically either from their human host's brain or from the device's own computer that is acting as a surrogate for the host's brain – to some organ, device, or system within or outside of the host's body for physical actuation.[17] Devices of this sort are already being used to fill a wide range of roles in treating diverse medical conditions and providing therapeutic benefits to many people around the world. For example, motor neuroprostheses are capable of detecting and interpreting their host's thoughts in order to allow the host to steer a wheelchair or guide a cursor around a computer screen.[18] They can provide life-altering benefits as the only means of communication with the outside world for locked-in patients who are completely paralyzed yet fully conscious, including those suffering from ALS, stroke, or traumatic brain injury.[19] They are also used to control internal bodily actions – for example, to restore bladder function after spinal cord injury, eliminate the need for an external ventilator in severely paralyzed individuals, or stimulate nerves that coordinate breathing and swallowing reflexes in order treat sleep apnea or facilitate swallowing after a stroke.[20] Motor neuroprostheses can also potentially be used to predict[21] or stop[22] epileptic seizures. They can

---

[16] For the possibility of visual cortical implants, see Thanos et al., "Implantable Visual Prostheses" (2007).

[17] See Lebedev (2014) and Patil & Turner, "The Development of Brain-Machine Interface Neuroprosthetic Devices" (2008).

[18] See Edlinger et al., "Brain Computer Interface" (2011); Lebedev (2014); Merkel et al., "Central Neural Prostheses" (2007); and Widge et al., "Direct Neural Control of Anatomically Correct Robotic Hands" (2010).

[19] See Donchin & Arbel, "P300 Based Brain Computer Interfaces: A Progress Report" (2009).

[20] See Taylor, "Functional Electrical Stimulation and Rehabilitation Applications of BCIs" (2008).

[21] For the use of EEG-based systems for this purpose, see Drongelen et al., "Seizure Prediction in Epilepsy" (2005).

[22] See Fountas & Smith, "A Novel Closed-Loop Stimulation System in the Control of Focal, Medically Refractory Epilepsy" (2007).

be used for functional electrical stimulation (FES) to restore muscle function-ality to individuals suffering from paralysis[23] (either as a permanent assistive technology or temporary rehabilitative tool[24]), to treat central hypoventila-tion syndrome,[25] and for neurally augmented sexual function (NASF) to re-store or improve sexual function in both male and female subjects.[26] Mean-while, the use of BCI devices for vagus nerve stimulation (VNS) is being ex-plored or considered to treat conditions such as Alzheimer's disease, anxiety disorders, bulimia, addictions, and narcolepsy.[27] Nontherapeutic applications of motor brain-computer interface (BCI) technologies have included, for ex-ample, the use of an EEG-based BCI to allow its human operator to drive a car in a 3D virtual reality environment.[28]

Some motor neuroprostheses are implanted in or interface with neurons in their host's brain, detecting neuronal activity that relates to a conscious volition or unconscious motor instruction and translating that activity into an output stimulus or signal produced by the device that activates or informs the functioning of transmission mechanisms that carry instructions to the motor plants or effectors that ultimately manifest the motor action. Other motor neuroprostheses directly perform the work of transmitting such in-struction-bearing stimuli to a motor plant or effector (in the human organ-ism, typically via a neuroeffector junction); still others receive and interpret such instructions and then execute the intended action through control of a motor plant, motor organ, or effector. Technologies that can be used to detect intent manifested within a human organism include electroencephalography (EEG), electrocorticography (ECoG), recordings of local field potentials (LFPs), and recordings of single-neuron action potentials,[29] as well as func-tional near infrared spectroscopy (fNIR).[30] Each technology has its unique

---

[23] See Durand et al., "Electrical Stimulation of the Neuromuscular System" (2005), and Moxon, "Neurorobotics" (2005).

[24] See Masani & Popovic, "Functional Electrical Stimulation in Rehabilitation and Neurorehabil-itation" (2011).

[25] See Taira & Hori, "Diaphragm Pacing with a Spinal Cord Stimulator: Current State and Future Directions" (2007).

[26] See Meloy, "Neurally Augmented Sexual Function" (2007).

[27] See Ansari et al., "Vagus Nerve Stimulation: Indications and Limitations" (2007).

[28] See Zhao et al., "EEG-Based Asynchronous BCI Control of a Car in 3D Virtual Reality Environ-ments" (2009).

[29] See Principe & McFarland, "BMI/BCI Modeling and Signal Processing" (2008).

[30] See Ayaz et al., "Assessment of Cognitive Neural Correlates for a Functional Near Infrared-Based Brain Computer Interface System" (2009).

strengths and weaknesses; for example, single-neuron recording is more invasive than EEG but less likely to be affected by artifacts from skin and muscle activity.[31]

3. Bidirectional sensorimotor neuroprostheses

Bidirectional sensorimotor neuroprostheses combine sensory and motor neuroprostheses in a single device that both provides sense data to the device's human host and receives instructions from the host that control the movement or other operation of the device. Some kinds of advanced prosthetic limbs are bidirectional sensorimotor neuroprosthetics: for example, an artificial hand may not only allow its human host to control the motion of the hand's fingers simply by willing such movements, but it may also provide the host with the ability to feel an object grasped within the hand and to sense how much pressure is being generated from the hand's contact with the object.[32]

Although most contemporary VR video game systems do not satisfy the definition of 'neuroprostheses' offered here (insofar as they do not directly integrate with a player's neural circuitry), systems that allow a player to control his or her action in a virtual game-world by motions of his or her real-world body (e.g., registered using motion-detecting sensors) and which then provide through the VR headset immediate visual and auditory feedback about the way in which the player's action has changed the game-world offer an example of the kind of intense biocybernetic feedback cycle that can be generated using bidirectional sensorimotor technologies.[33]

4. Cognitive neuroprostheses

A cognitive neuroprosthetic device participates in or supplements processes that are internal to the mind of its human host and which do not directly involve either sensory or motor organs (although the processes may receive input from or transmit output to such organs). Such neuroprosthetic devices may participate in cognitive processes and phenomena such as

---

[31] See Miller & Ojemann, "A Simple, Spectral-Change Based, Electrocorticographic Brain–Computer Interface" (2009).

[32] See Hoffmann & Micera, "Introduction to Neuroprosthetics" (2011), pp. 792-93.

[33] See Gladden, "Cybershells, Shapeshifting, and Neuroprosthetics: Video Games as Tools for Posthuman 'Body Schema (Re)Engineering'" (2015).

memory, imagination,[34] emotion,[35] belief, identity,[36] agency, attentiveness, consciousness,[37] and conscience.

The development of such technologies for use in human beings is still in its earliest stages. Although not the primary purpose for which the devices were designed, effects relating to creativity and one's sense of authenticity and agency have been reported in patients utilizing neuroprosthetic devices for deep brain stimulation.[38] Mnemoprosthetic devices that allow the creation or alteration of memories by manipulating the brain's natural mechanisms for the storage of memories have been experimentally tested in mice[39] and in principle could potentially be employed with the human brain, as well. However, such technologies currently fall far short of allowing the implantation of complex, content-rich memories into a mind or allowing the precise and detailed editing of existing memories.[40] Indeed, deep mysteries exist regarding the mechanisms by which long-term memories are created, stored, and retrieved in the human mind, and divergent theories have been proposed to explain the functioning of such systems.[41] As neuroscience continues to advance and competing theories are either confirmed or rejected, we will learn more about the kinds of cognitive neuroprosthetic devices that theoretically can or cannot be created and successfully integrated into the neural circuitry and functioning of a human mind. (And conversely, the ability or inability to

---

[34] See Cosgrove, "Session 6: Neuroscience, brain, and behavior V: Deep brain stimulation" (2004), and Gasson, "Human ICT Implants: From Restorative Application to Human Enhancement" (2012).

[35] For the possibility of developing emotional neuroprostheses, see Soussou & Berger, "Cognitive and Emotional Neuroprostheses" (2008); Hatfield et al., "Brain Processes and Neurofeedback for Performance Enhancement of Precision Motor Behavior" (2009); Kraemer, "Me, Myself and My Brain Implant: Deep Brain Stimulation Raises Questions of Personal Authenticity and Alienation" (2011); and McGee, "Bioelectronics and Implanted Devices" (2008), p. 217.

[36] See Kraemer (2011) and Van den Berg, "Pieces of Me: On Identity and Information and Communications Technology Implants" (2012).

[37] For the possibility of neuroprosthetic devices relating to sleep, see Claussen & Hofmann, "Sleep, Neuroengineering and Dynamics" (2012), and Kourany, "Human Enhancement: Making the Debate More Productive" (2013), pp. 992-93.

[38] See Kraemer (2011).

[39] See Han et al., "Selective Erasure of a Fear Memory" (2009); Josselyn, "Continuing the Search for the Engram: Examining the Mechanism of Fear Memories" (2010); and Ramirez et al., "Creating a False Memory in the Hippocampus" (2013).

[40] For questions about the extent to which technological devices that directly store memories can ever become a part of the human mind, see Clowes, "The Cognitive Integration of E-Memory" (2013).

[41] See, for example, Dudai, "The Neurobiology of Consolidations, Or, How Stable Is the Engram?" (2004).

successfully develop and implement particular kinds of neuroprosthetic devices may shed light on whether particular proposed brain theories are correct or incorrect.) For example, if holographic brain models[42] were found to be correct, it might largely rule out the possibility of constructing neuroprosthetic devices that can create or alter a complex long-term memory simply by manipulating a modest number of neurons in a particular region of the brain.

B. Neuroprosthetic devices categorized by purpose: therapy vs. enhancement

In addition to categorizing neuroprosthetic devices according to their function (i.e., as sensory, motor, bidirectional, or cognitive), such devices may also be categorized according to their purpose. For example, some neuroprosthetic devices are used for purposes of therapeutic **restoration**, to restore abilities that have been lost by a human being due to illness or injury. Other neuroprosthetic devices do not directly treat a medical condition but are instead used for purposes of **diagnosis**, to gather information about the condition of their human host and allow medical decisions to be made. Still other neuroprosthetic devices may be used for purposes of **identification**, to verify the identity of the device's human host, allow his or her whereabouts or activities to be tracked, or allow him or her access to some restricted area or resource.[43] Finally, some neuroprosthetic devices are designed for purposes of human **enhancement**: such devices augment, modify, or replace the sensory, motor, or cognitive abilities of their human host, allowing him or her to experience phenomena or perform actions that are not possible for the minds and bodies of natural, unmodified human beings.[44]

C. Neuroprosthetic devices categorized by physical location: implant vs. prosthesis

Neuroprosthetic devices may alternatively be categorized according to their relationship with the body of their human host.[45] In this text, we use the word 'implant' to describe a neuroprosthetic device that is surgically inserted into the body of its human host and which remains within the host's body

---

[42] Such models have been described, e.g., in Longuet-Higgins, "Holographic Model of Temporal Recall" (1968); Westlake, "The possibilities of neural holographic processes within the brain" (1970); Pribram, "Prolegomenon for a Holonomic Brain Theory" (1990); and Pribram & Meade, "Conscious Awareness: Processing in the Synaptodendritic Web – The Correlation of Neuron Density with Brain Size" (1999). An overview of conventional contemporary models of long-term memory is found in Rutherford et al., "Long-Term Memory: Encoding to Retrieval" (2012).

[43] The term 'identification' has been used here in a loose sense; from the perspective of information security, what has just been described as 'identification' actually involves identification, authentication, and authorization.

[44] See Gasson (2012), p. 25.

[45] See Gasson (2012), p. 14.

during the device's operation. Devices that are introduced into the body of their human host by nonsurgical means (such as nanorobots that are orally ingested) would not be 'implants' in this sense, even if they establish a permanent connection with particular neurons after their entry into their host's body; such technologies could be described more broadly as '**endosomatic**' devices or systems that are housed within their host's body but are not surgically implanted. Neuroprosthetic devices formed of biological components that are grown or cultivated *in situ* within their host's body would be another example of such endosomatic systems that are not, strictly speaking, implants.

Meanwhile, we can define a neurocybernetic '**prosthesis**' as a device that is integrated into the neural circuitry of its human host but which is not completely contained within the host's body; it instead forms part of the exterior surface or boundary of the body and extends the body outward into the surrounding environment. It is possible for a single device to be both a neuroprosthetic implant and a prosthesis: for example, an artificial eye that has been surgically implanted but which (at least, when the eyelid is open) forms part of the body's exterior surface and a portion of its physical interface with the external environment. It is also possible for an implant and prosthesis to work together closely as part of a larger system. For example, an individual who has lost an arm due to injury may now possess a permanent implant located in the shoulder area that is integrated with the sensory and motor nerves that previously innervated the arm. If that implant contains an external socket that allows different robotic arms to be attached to it and controlled by the device's host (or which perhaps even allows different kinds of robotic limbs and manipulators to be swapped in and out of the socket), then the socket itself would be considered an implant, and a robotic arm capable of connecting with the socket (and, through it, becoming indirectly integrated into the neural circuitry of the device's human host) would be considered a prosthesis.

Note that some other texts that focus specifically on brain-computer interfaces (BCIs) may use terms such as 'invasive,' 'partially invasive,' and 'noninvasive' to refer to a device's physical relationship to the *brain* of its human host rather than its relationship to the host's body as a whole. According to such definitions, a device could be wholly contained within the body of its human host but would be classified as 'noninvasive' if it were implanted in, say, the host's abdomen rather than the gray matter of his or her brain.[46] As defined in this text, a neuroprosthetic device must be integrated into the 'neural circuitry' of its human host, but this does not necessarily require a

---

[46] See Gasson (2012), p. 14, and Panoulas et al., "Brain-Computer Interface (BCI): Types, Processing Perspectives and Applications" (2010).

connection to interneurons contained within the *brain*; a neuroprosthetic device could be located elsewhere in the body and possess a physical interface with afferent or efferent neurons in that location. In the context of this book, 'invasive' is best used to refer to neuroprosthetic devices that are endosomatic or fully contained within the human body of their host; 'noninvasive' neuroprosthetic devices would be those that have no physical components contained within the body of their human host (such devices might include wearable neuroprostheses that rest on the external surface of the body and which communicate with neurons via signals transmitted through the skin, or even devices that can communicate with neurons at a greater distance through the generation and detection of electromagnetic fields or radiation); and 'semi-invasive' neuroprosthetic devices would be those that have components that are external to (and perhaps not even physically connected to) their host's body but which simultaneously possess some components that must be introduced into the body of their human host (such as electronic components that must be inserted into the ear canal or through a permanent port installed in the body via a surgically created stoma, or biochemical agents that must be introduced into the host's bloodstream).

D. Neuroprosthetic devices categorized by agency: active vs. passive

With regard to their interaction with the biological structures and processes of their human host, some neuroprosthetic devices may be considered 'active,' insofar as they possess an internal computer or other mechanism (e.g., a transmitter that allows the device to receive instructions from an external system) that governs the device's behavior and allows the device to proactively undertake actions and to determine how it will respond to stimuli received from its human host or the external environment. A 'passive' neuroprosthetic device, on the other hand, is essentially an inert tool that lacks its own centralized internal control mechanism and whose behavior is controlled by the biological processes of and input supplied by the device's human host.[47]

An artificial eye that uses its built-in video camera to register light from the external environment, utilizes its internal computer to process those incoming signals and convert them into a pattern of stimuli, and then stimulates retinal ganglion cells according to that pattern would be an active neuroprosthetic device; its internal computer governs its behavior, and if an adversary were able to access and compromise the computer, he or she could

---

[47] For one approach to classifying information and communications technology (ICT) implants as 'active' or 'passive' with regard for their functionality, see Roosendaal, "Implants and Human Rights, in Particular Bodily Integrity" (2012).

potentially use the device to supply its host with manipulated or even entirely fabricated visual data.[48]

On the other hand, an example of a passive implant would be an array of synthetic biomimetic physical neurons that is implanted into its host's brain to replace a group of natural biological neurons that had been destroyed through illness or injury. Although each individual synthetic neuron may possess a limited form of agency and control over its own actions – insofar as it possesses mechanisms that determine how it will react to particular stimuli – the device as a whole possesses no centralized control mechanism and is, in essence, an empty scaffolding that cannot fill itself with information or decide to take action. The natural biological neurons that are connected to the implant may eventually begin to 'use' it by supplying stimuli to it and incorporating it into their network of activity and information storage, but such action cannot be forced or compelled by the implant itself.

Attacks against active vs. passive neuroprostheses

Note that if a neuroprosthetic device is controlled by an internal computer that possesses its own memory, processor, and input/output mechanisms and which runs its own operating system (and potentially additional specialized software), the device is almost certainly an 'active' one, even if the intended purpose of the device is simply to detect the wishes and volitions of its human host and then to execute them. Although such a device may typically operate in a way that creates the *appearance* that it is strictly passive, an adversary who gained access to the device's computer and compromised its hardware or software could use the device as (or turn the device into) an active agent that behaves in ways that were not at all requested or desired by the device's human host. On the other hand, a purely passive neuroprosthetic device could not be directly hijacked by an adversary and utilized to perform certain actions or behaviors, because the device itself has no internal control mechanism that can be commandeered; the only way that an adversary could indirectly dictate the actions of a passive neuroprosthetic device (without radically reengineering the device itself) would be to control the biological structures or processes of the device's human host that interact with the device, causing them to externally stimulate the device in ways that would produce a particular response.

E. Implantable computers vs. neuroprosthetic devices

Not all implantable computers are neuroprosthetic devices: it is possible to have a miniaturized computer (e.g., as part of an active RFID transponder) that is implanted within a human being's body but which has no interface or

---

[48] Regarding such possibilities of neuroprostheses being used to provide false data or information to their hosts or users, see McGee (2008), p. 221.

interaction with the person's neural circuitry. Conversely, not every neuro-prosthetic device is (or contains) an 'implantable' computer: for example, the external portion of a prosthetic arm may contain a highly sophisticated computer that is integrated into the neural circuitry of its human host through a stable physical connection and interaction with nerves in the person's shoulder, however the computer would be considered part of a 'prosthesis' rather than an 'implant.'

II. Expected developments in neuroprosthetics: toward posthuman enhancement

The kinds of neuroprosthetic devices that are currently in widespread use have typically been designed to serve a restorative or therapeutic medical purpose – for example, to treat a particular illness or restore some sensory, motor, or cognitive ability that their user has lost as a result of illness or injury. It is expected, though, that future generations of neuroprostheses will increasingly be designed not to restore some ordinary human capacity that is absent but to enhance their user's physical or intellectual capacities by providing abilities that exceed or differ from what is naturally possible for human beings.[49] The potential use of such technologies for physical and cognitive enhancement is expected to expand the market for neuroprostheses and implantable computers to reach new audiences well beyond the limited segment of the population that currently relies on them to treat medical conditions.[50]

Researchers expect that future versions of sensory neuroprostheses such as retinal implants may give human beings the capacity to experience their environments in dramatically new ways, for example through the use of telescopic or night vision[51] or by using a form of augmented reality that overlays actual sense data provided by the environment with supplemental information received or generated by a neuroprosthetic device's computer.[52] Some researchers envision the development of devices that resemble more sophisticated forms of retinal and cochlear implants that can record all of a person's audiovisual experiences for later playback on demand, effectively granting

---

[49] See Gasson (2008); Gasson et al. (2012); McGee (2008); and Merkel et al. (2007).

[50] See McGee (2008) and Gasson et al. (2012).

[51] See Gasson et al. (2012) and Merkel et al. (2007).

[52] See Koops & Leenes, "Cheating with Implants: Implications of the Hidden Information Advantage of Bionic Ears and Eyes" (2012).

the person perfect audiovisual memory[53] and potentially allowing the individual to share his or her sensory experiences with others (e.g., through automatic upload to a streaming website).

Building on successful experiments with implanting artificial memories in mice, other researchers have envisioned the possibility of a person being able to regularly download new content onto a memory chip implanted in his or her brain, thereby instantaneously gaining access to new knowledge or skills.[54] Even more futuristic scenarios envisioned by scholars include the development of a 'knowledge pill' that can be ingested and whose contents – perhaps a swarm of web-enabled nanorobots[55] – travel to the brain, where they modify or stimulate neurons to create engrams containing particular memories.[56] Another potentially revolutionary technological advancement is the ongoing development of brain-machine-brain interfaces[57] that may eventually allow direct and instantaneous communication between two human brains physically located thousands of miles apart.

---

[53] See Merkel et al. (2007) and Robinett, "The consequences of fully understanding the brain" (2002).

[54] See McGee (2008).

[55] See Pearce, "The Biointelligence Explosion" (2012).

[56] See Spohrer, "NBICS (Nano-Bio-Info-Cogno-Socio) Convergence to Improve Human Performance: Opportunities and Challenges" (2002).

[57] See Rao et al., "A direct brain-to-brain interface in humans" (2014). Existing experimental technologies of this sort are sometimes described as 'brain-brain interfaces' (BBIs), although we would argue that such terminology is somewhat misleading; it would be more appropriate to describe the system as a 'brain-machine-brain interface' (BMBI) or 'brain-computer-brain-interface' (BCBI). If one were allowed to describe as a 'brain-brain interface' a system that actually interposes between the two brains some complex technological device that enables and mediates their communication, then traditional technologies such as telephones and even books could similarly be described as 'brain-brain interfaces' with just as much legitimacy. It can be argued that a true 'brain-brain interface' would instead be one in which the communication between the two brains does not rely on any 'external' device or system; rather, the means of communication between the two brains would be contained within and fully integrated into one or both of the brains themselves. An electronic transmitter that is permanently implanted within a host's brain and which harvests energy from the brain itself and allows the brain to communicate with other brains possessing similar devices could conceivably be described as a 'brain-brain interface.' A clearer example would be that of a prosthesis composed of biological material that is either implanted into or grown or assembled within a brain, and which through its unique organic design is capable of generating and detecting radio frequency transmissions, light, electromagnetic fields, ultrasonic waves, or other phenomena that are detectable at a distance. If two brains possessing such prostheses were able to communicate with one another by means of the devices, this could be understood as an example of a 'brain-brain interface,' even if the devices in fact were reliant on a medium (such as that of the atmosphere) for transmission of their signals. In its functioning, such a system would approach traditional definitions of 'telepathy.'

A. Early adopters of neuroprosthetic devices for posthuman enhancement

One group of potential 'early adopters' of neuroprosthetic devices designed for human enhancement includes military forces, intelligence agencies, police forces, and other government agencies that may use such technologies to enhance the capacities of their personnel to engage in conventional combat operations, cyberwarfare, and the gathering and analysis of intelligence.[58] Another potential group of early adopters of such technologies includes hardcore computer gamers (including professional competitive gamers) who wish to experience more sophisticated and immersive forms of sensorimotor interaction with game-worlds and cybernetic interaction with their fellow gamers than can be provided by external virtual reality systems.

B. The meaning of 'advanced' neuroprosthetics

This text addresses the necessity of and practices for ensuring information security for advanced neuroprosthetic devices. By 'advanced,' we mean that this book considers all types of neuroprosthetic devices whose future development is anticipated and not simply those kinds that already are in widespread use among human beings (like cochlear implants), are undergoing testing for therapeutic use in human beings (like retinal prostheses with limited visual resolution), or which are currently being tested in animals but could potentially be adapted someday for use in human beings (like some kinds of mnemoprostheses designed to create or alter particular memories).

Many of the IMDs that are currently in use around the world – especially those that were implanted years or even decades ago – present both an advantage and a unique challenge from the perspective of information security, insofar as their internal computers are severely constrained in their capacities and functionality; this may prevent one from applying conventional InfoSec mechanisms and software that are commonly employed with more powerful computers (e.g., those found in desktop computers or smartphones) while simultaneously shielding the devices from attacks to which only more powerful conventional computers and operating systems may be vulnerable. Looking ahead to the future, though, we can anticipate the need to provide information security to implanted neuroprosthetic devices that differ radically from today's best desktop computers not in being much less powerful than they are but in being much *more* so – or in utilizing exotic hardware and software platforms (such as biomolecular computing) that have little in common with today's computers designed for general office or home use.

---

[58] On potential military use of neuroprosthetic devices, see Schermer, "The Mind and the Machine. On the Conceptual and Moral Implications of Brain-Machine Interaction" (2009), and Brunner & Schalk, "Brain-Computer Interaction" (2009).

While this text considers all such devices that are currently in widespread use or are undergoing testing, the scope of the book is broader: it also addresses the information security needs of those more advanced kinds of neuroprosthetic devices (such as artificial eyes possessing human-like visual resolution) that scientists, engineers, and entrepreneurs have declared their intention to create and are actively working to bring to market, as well as more sophisticated neuroprosthetic devices whose eventual development is expected by researchers and professional futurists and whose legal, ethical, political, economic, cultural and technological implications are already being debated by the proponents and critics of such technologies.

Among such potential future neuroprosthetic technologies are ones that may allow human beings to acquire new sensory capacities, adopt radically nonhuman bodies, inhabit virtual worlds in which different laws of physics and biology hold sway, and directly link their minds with one another or with artificial intelligences to create new kinds of communal thought and agency.[59] Although today there is not yet a widespread practical necessity to *implement* InfoSec mechanisms and procedures for such systems, it is important to begin developing the theoretical, conceptual, and organizational frameworks that will be needed to promote the information security of systems utilizing such technologies – especially insofar as the formulation of sound information security frameworks can aid those individuals and organizations that are actively pursuing the development of such technologies, to help ensure that they are designed and eventually deployed in ways that will advance rather than undermine essential aims such as human authenticity, agency, and full human development.[60]

---

[59] See Merkel et al. (2007); Gladden, "Cybershells, Shapeshifting, and Neuroprosthetics" (2015); and Gladden, "Enterprise Architecture for Neurocybernetically Augmented Organizational Systems" (2016).

[60] See Gladden, "Neural Implants as Gateways to Digital-Physical Ecosystems and Posthuman Socioeconomic Interaction" (2016).

Chapter Two

An Introduction to Information Security in the Context of Advanced Neuroprosthetics

Abstract. This chapter provides an introduction to basic principles and practices within the complex, diverse, and dynamic field of information security. Concepts such as the CIA Triad; the nature of administrative, logical, and physical security controls; the role of access controls in performing user authentication and authorization; and the differences between vulnerabilities, threats, and risks are all discussed. Also highlighted are the different forms that information security can take when pursued by a large organization as opposed to, say, the individual user of a consumer electronics device.

I. Institutional processes and personal practices

Information security (also known as InfoSec) has been defined as "The protection of information and information systems from unauthorized access, use, disclosure, disruption, modification, or destruction in order to provide confidentiality, integrity, and availability."[1] The work of attempting to provide such protection is a never-ending process; information security is not a static state that can be achieved once and for all but is rather a goal to be continuously pursued. While large organizations are able to implement sophisticated information security plans that are developed and executed by dedicated specialists, even individual consumers and users of information technology instinctively utilize many information security practices in an effort to keep sensitive information private and secure. In this chapter we briefly describe some of the foundational concepts of information security that will be explored in greater depth throughout this book.

---

[1] This definition is formalized in 44 U.S.C., Sec. 3542 and cited, e.g., in *NIST Special Publication 800-53, Revision 4: Security and Privacy Controls for Federal Information Systems and Organizations* (2013), p. B–10.

II. The CIA Triad

A key principle of information security is that of ensuring the **confidentiality**, **integrity**, and **availability** of information for authorized use by approved users. The original 'CIA Triad' of information security objectives[2] has been expanded by various researchers and practitioners through the addition of other proposed objectives,[3] but it remains the most concise and universally recognized summary of the goals that the field of information security seeks to achieve. We will consider the relationship of the CIA Triad and other information security objectives to neuroprosthetic devices in detail in Chapter Four.

It is essential to note that as it relates to advanced neuroprostheses, information security is not only concerned with maintaining the confidentiality, integrity, and availability of information contained within a neuroprosthetic device itself (such as proprietary software used to control the device). Insofar as a neuroprosthetic device is linked to the neural circuitry of its human host, an adversary who gains unauthorized access to the device could compromise the confidentiality of information relating to the internal biological processes and medical status of its host – and could potentially even detect the contents of cognitive processes such as thoughts, fears, imaginings, emotions, memories, and beliefs. Moreover, the integrity and availability of information stored or existing within the mind of a human host could also be at risk if an adversary could use a neuroprosthetic device to alter, manipulate, or damage the biological structures, biochemical and bioelectrical activity, and patterns of information stored within the host's network of neurons.[4] The designers and operators of neuroprosthetic systems must also take into account the fact that a device such as an artificial eye could conceivably be illicitly accessed by an adversary not to gather information about the host in whose body the device is implanted but about some unrelated party who is the actual target of the adversary's unauthorized surveillance and who might only coincidentally happen to be standing near the device's host in a given moment. Thus the presence of advanced neuroprosthetic devices within a human society can potentially impact the confidentiality, integrity, and availability of information for individuals beyond the limited number of persons who personally possess such devices.

---

[2] See Rao & Nayak, *The InfoSec Handbook* (2014), pp. 49-53.

[3] See Parker, "Toward a New Framework for Information Security" (2002), and Parker, "Our Excessively Simplistic Information Security Model and How to Fix It" (2010).

[4] See Merkel et al., "Central Neural Prostheses" (2007), and Gladden, "Neural Implants as Gateways to Digital-Physical Ecosystems and Posthuman Socioeconomic Interaction" (2016).

III. Security controls

One of the primary means of protecting information is to design and implement effective **security controls**. A security control can be understood as "A safeguard or countermeasure prescribed for an information system or an organization designed to protect the confidentiality, integrity, and availability of its information and to meet a set of defined security requirements."[5] Such controls can be either **administrative** (i.e., consisting of organizational policies and procedures), **physical** (i.e., created by physical walls and barriers, motion-detecting alarm systems, electric fences, security guards, or the physical isolation of a computer from any wired network connections), or **logical** (i.e., enforced through software or other electronic or computerized decision-making).[6] Together, administrative and physical controls can be understood as 'non-computing security methods,' which "are security safeguards which do not use the hardware, software, and firmware of the IT."[7]

A. An example of administrative, logical, and physical controls

Consider, for a moment, an advanced cochlear implant that not only presents its human host with high-fidelity live audio received from the external environment but which also: 1) records all of the auditory data that it receives to an internal memory card that the device's host can access at any time through an act of will to 'play back' previous auditory experiences, and 2) wirelessly transmits a live stream of the recorded audio to a cloud-based storage system to create a permanent external backup of all the host's auditory experiences.

Such a device would gather – and have the potential to gather – vast and diverse quantities of data relating to the device's human host, other human beings or technological systems that interact with the host, and other human beings or systems that simply happen to be near the host. The device would likely record not only all of the host's personal and professional conversations (as well as all of the other conversations taking place in, say, a crowded restaurant in which the host was dining), but also all of the auditory contents of all the music, films, theatrical performances, and computer games that the host experiences, as well as the sound of the host's heartbeat and breathing (which can be used to infer details about his or her physical activities and health status).

In the case of such a device, examples of physical controls might include designing, constructing, and implanting the device in such a way that it has

---

[5] *NIST SP 800-53* (2013), p. B–21.

[6] Rao & Nayak (2014), pp. 66-69.

[7] *NIST Special Publication 800-33: Underlying Technical Models for Information Technology Security* (2001), p. 21.

no external ports or sockets that are accessible from the exterior of its host's body (and which adversaries could potentially illicitly access) or adding a tamper-proof seal that prevents physical access to the device's internal components. Within an organizational setting, physical controls might also include soundproofed walls and secured doors that prevent individuals who possess such devices from entering parts of the building in which the devices could be illicitly used to record sensitive conversations.

Logical (or 'technical') controls are those grounded in the ability of software to analyze decisions and take particular actions according to specified rules. For example, the advanced cochlear implant's internal computer might be programmed with filters that prevent the recording of sounds of particular volumes and frequencies (e.g., to block the recording of its owner's heartbeat) or even to identify and prevent the recording and playback of particular kinds of sounds (e.g., commercially released music) representing intellectual property whose use is legally restricted. The device may also use logical controls to allow its human host and authorized maintenance or medical personnel to wirelessly log into the device's command console and reconfigure its settings while blocking unauthorized users from gaining such access. An organization may also deploy software within the ubiquitous computing environment of its R&D facility that detects the presence of visitors who possess such advanced cochlear implants and broadcasts white noise in targeted locations to prevent such devices from recording sensitive conversations among the facility's staff. While the sound waves of the white noise itself might constitute a physical control, the software that analyzes data to determine the presence of visitors within the building and decides whether, when, and how to broadcast such sounds would constitute a logical control (which could potentially be disabled or evaded through logical means).

Administrative controls are those grounded in policies and procedures that are designed to be carried out by human beings.[8] The extent to which

---

[8] Although there may be circumstances in which an artificial agent or system participates in the enforcement of administrative controls, the role of such a software-based technological system can often be better understood as constituting a logical rather than administrative control. It could be appropriate to speak of an artificial system as managing the implementation of administrative controls in a human-like sense if the system possessed a kind and degree of artificial intelligence (e.g., produced by an advanced physical artificial neural network) such that its behaviors were not strictly determined by the code contained in a particular computer program but instead resulted from the system's application of its own reasoning, memories, imagination, values, conscience, and judgment to arrive at decisions that were not necessarily entirely predictable. Such an artificially intelligent system would – like human beings – be capable of either strictly enforcing administrative controls, allowing periodic (and unauthorized) exceptions to the administrative controls on the basis of the agent's own values and personal experience, shirk-

they are successfully enforced depends on many factors, including the competence, commitment, training, and motivations of the individuals charged with enforcing them. In the case of our advanced cochlear implant, the host of such a device might develop his or her own personal administrative controls or policies relating to its use. For example, perhaps the person has decided that he or she will always disable the device's recording function (while still allowing himself or herself, however, to experience a live stream of the auditory data received by the device) whenever engaging in certain kinds of sensitive conversations with particular individuals or while asleep during the night. The company that designs and manufactures the device, meanwhile, might develop its own policies and procedures that determine how different kinds of authorized maintenance and emergency medical personnel around the world can apply for and receive a special kind of certification and an access code that allows them to log into the command console of any such device.[9] Insofar as it is legally and ethically permissible, an organization may also develop policies that require all of its employees who possess or acquire such an advanced cochlear implant to notify the organization of this fact and to follow specified practices that restrict them from entering certain areas of the organization's facility in which they could purposefully or unintentionally record auditory data containing highly sensitive organizational information.

We will consider the use of such administrative, logical, and physical controls to provide information security for neuroprosthetic devices and systems in greater detail in Chapters Six, Seven, and Eight (where they will, however, be categorized primarily according to their role as preventive, detective, or corrective and compensating controls).

B. Overlap of different kinds of security controls

In principle, it should be possible to categorize security control systems according to whether they fill a role of authentication or authorization or whether they are administrative, logical, or physical. In practice, though, there is significant overlap among categories. For example, a company's server might be protected inside a dedicated server room whose door is secured with a tumbler lock that requires a traditional physical key to open. The fact that a user must possess a physical key in order to open the door

---

ing one's responsibilities and failing to enforce administrative controls out of laziness or resentment, failing to enforce administrative controls because they were not properly understood, or periodically failing to enforce particular administrative controls because – in the agent's best judgment – resources or attention need to be allocated to other, more important priorities, instead.

[9] For a discussion of such certificate schemes for implantable medical devices, see, e.g., Cho & Lee, "Biometric Based Secure Communications without Pre-Deployed Key for Biosensor Implanted in Body Sensor Networks" (2012), and Freudenthal et al., "Practical techniques for limiting disclosure of RF-equipped medical devices" (2007).

makes this system a physical access control. However, if, for example, the company has effective policies and procedures in place to govern the posses-sion and use of the key (e.g., only two members of the IT staff have a copy of the key, or the key is stored in a safe in a different part of the company's fa-cility and can only be temporarily checked out by IT personnel when needed), then it is also part of an administrative control system. Moreover, if the phys-ical key also contains an embedded RFID chip that must be detected by the lock's electronic RFID reader before the lock will open (in order to prevent unauthorized copies of the key from opening the lock), then the key would also be part of a logical access control system. In addition, the key not only serves a purpose of authenticating its user (since presumably only one of the server room's intended users would have a copy of the key), but it also serves the purpose of authorization by allowing the authenticated user to open the door, enter the room, and access the resources inside, while the lack of a key prevents unauthenticated parties from entering the room.

Similarly, a closed-circuit TV camera that monitors the hallway leading to the server room is a physical control, insofar as it creates a physical obstacle that an unauthorized party seeking access to the server room might try to avoid. If the live video is being processed by a software program that contin-ually scans the images and alerts a human security staff member when par-ticular phenomena are detected in the video (e.g., images of a human being approaching the server room door), this would constitute a logical control. Finally, if the human security staff member – upon seeing the live video of someone approaching the server room door – follows a particular set of guide-lines to decide whether he or she should go and investigate the situation or simply ignore it, this would be an administrative control.

IV. Authentication and authorization

In order to assure that information is readily available for use by its in-tended users but not available to any other parties, there must be some sys-tem or mechanisms in place that can either block or unlock access to the in-formation, depending on the circumstances of who is seeking the information and for what purpose. Access controls are systems or mechanisms of this sort that are designed to "Enable authorized use of a resource while preventing unauthorized use or use in an unauthorized manner."[10] Rao and Nayak note that such access controls are "considered the most important aspect of infor-mation security."[11]

---

[10] *NIST SP 800-33* (2001), p. 20.
[11] Rao & Nayak (2014), p. 63.

An access control must perform two key functions.[12] First, it carries out the step of **authentication**, or verifying the identity of the entity that is attempting to access the protected information – whether that entity be a human being, device, or software process.[13] Second, the access control carries out the step of **authorization**, or granting the user access to those portions of the information system and the information contained within it that the user is authorized to access[14] (which, in the case of unauthorized users, may be no information at all).

V. Vulnerabilities, threats, and risks

Within the field of information security, words such as 'vulnerability,' 'threat,' and 'risk' are often employed to convey specific technical meanings.[15]

A **vulnerability**, for example, can be understood as "A weakness in system security procedures, design, implementation, internal controls, etc., that could be accidentally triggered or intentionally exploited and result in a violation of the system's security policy."[16] In the case of neuroprosthetic devices, we might say, for example, that a robotic prosthetic arm displays a significant vulnerability if it possesses an exposed USB port that can be used to access the device's internal computer. Similarly, an artificial eye displays a vulnerability if it has been designed in such a way that a beam of light directed at the eye with a particular frequency and intensity will automatically cause the artificial eye's internal computer to reboot.[17] Note that while such characteristics might be considered 'vulnerabilities' from the perspective of information security, from a functional and operational perspective they might simultaneously be considered essential features that are necessary to ensuring the

---

[12] Rao & Nayak (2014), p. 63.

[13] *NIST SP 800-33* (2001), p. 20.

[14] Rao & Nayak (2014), pp. 62-76.

[15] Throughout this text other words such as 'danger' and 'hazard' may sometimes be used; these terms are employed more generally and without a specific technical meaning. Depending on the context, they may refer to something that is a vulnerability, threat source, threat, risk, or some combination of these.

[16] *NIST SP 800-33* (2001), p. 18.

[17] The hypothetical situation of a poorly designed prosthetic eye whose internal computer can be disabled if the eye is presented with a particular pattern of flashing lights is raised in Hansen & Hansen, "A Taxonomy of Vulnerabilities in Implantable Medical Devices" (2010). In the example that we present here, the behavior of our hypothesized artificial eye is not a bug or flaw that results from poor design (and perhaps has not yet even been discovered by the device's designer or users) but rather a feature that has been intentionally added by the device's designer, e.g., to facilitate diagnostic, maintenance, or emergency control activities that may periodically need to be performed by maintenance or medical personnel.

effective functioning of the device and thus the health and well-being of its human host.

In themselves, the vulnerabilities displayed by a neuroprosthetic device do not cause a loss of information security for the device's host or operator; such harm results only when a particular **threat source** interacts with the device in such a way that – because of the existence of the vulnerability – compromises the confidentiality, integrity, or availability of information. A threat source may be an agent such as a human adversary or computer virus that purposefully targets and exploits the vulnerability (in which case the threat-source's action can be considered an 'attack'), or a threat source could be some unintentional and purely accidental occurrence within a neuroprosthetic device, its human host, or the surrounding environment that triggers the effects of the vulnerability.[18] In the case of the robotic prosthetic arm displaying the vulnerability of an exposed USB port, a threat source could be a human adversary with a USB flash drive containing malicious code that will hijack control of the arm and cause it to behave in a particular way after the flash drive is plugged into the port.[19] In the case of the artificial eye that automatically reboots when exposed to a particular frequency and intensity of light, a threat source could be a computer monitor on which a movie is being played that causes light of various frequencies and intensities to be emitted from the screen.

Similarly, we can understand a **threat** as the possibility that some threat source will either intentionally attempt to exploit a vulnerability (as in the case of a cyberattack) or will unintentionally trigger the vulnerability's effects (as in the case of some random environmental phenomenon).[20] In the case of the robotic prosthetic arm, a threat exists if there are adversarial agents in the environment attempting to exploit the device's vulnerability, or if there is a possibility that such threat sources may exist, or if other threat sources exist (such as USB flash drives containing executable code that is not intentionally malicious but which – if inserted into the device's USB port – would cause the device's internal computer to crash). In the case of the artificial eye, a threat exists if there are (or, as far as one knows, may be) light sources of the relevant frequency and intensity in areas that the device's human host might enter.

---

[18] *NIST SP 800-33* (2001), p. 18. In the case of implantable medical devices, the 'compromised vulnerability' through which a threat acts is also known as a 'threat vector' or 'etiological agent'; see Hansen & Hansen (2010).

[19] For the possibility of an adversary gaining control of a prosthetic limb, see Denning et al., "Neurosecurity: Security and Privacy for Neural Devices" (2009).

[20] *NIST SP 800-33* (2001), p. 18.

With regard to an advanced neuroprosthetic device, the **risk** arising from a particular vulnerability and set of threats is an impact borne by the device's human host, operator, and other individuals or organizations responsible for the device's development and use.[21] The risk is influenced by the probability that a threat source will intentionally attack the device's vulnerability or unintentionally trigger its effects as well as by the nature and extent of the harm that would occur as a result.

Best practice in the field of information security requires individuals and organizations to proactively address the risks that they face, which is done by developing and implementing a program of **risk management**. The first step in such a program is to determine the scope, nature, and context of the activities to be considered.[22] In the case of an organization that designs and manufactures neuroprosthetic devices, it may focus on the need to provide information security for the sensitive and valuable proprietary research and design data developed by its laboratories and engineers during the R&D stage as well as building into the devices basic systems to protect the information security of their hosts and end users. Meanwhile, an individual user who has purchased a neuroprosthesis as a consumer electronics device will presumably not have the desire or resources to monitor whether the device's manufacturer is able to keep confidential its internal research data; instead, he or she may focus on the information security risks arising from his or her intended use of the device in a very specific situation (e.g., to enhance his or her workplace performance or treat the symptoms of a particular neurological condition). Having determined the risk context, an individual or organization then conducts a **risk assessment** to identify specific vulnerabilities, threat sources, threats, and risks that are relevant, given the chosen risk context.[23]

Each individual and organization must determine how much risk it is willing to bear, taking all relevant elements into consideration. In some cases, **risk mitigation** can be employed to reduce the risk associated with a neuroprosthetic device to a level that the individuals or organizations involved find acceptable. Risk mitigation attempts to reduce a risk and its impact by seeking to either eliminate a vulnerability at its source, add protections that prevent the vulnerability from being intentionally exploited or accidentally triggered, or reduce and contain the harm that will result from exploitation or triggering of the vulnerability.[24] In the case of the robotic prosthetic arm described above, risk mitigation might involve removing or disabling the USB port, adding a cosmetic cover that conceals its existence, or installing security software

---

[21] *NIST SP 800-33* (2001), p. 18.
[22] *NIST SP 800-33* (2001), p. 19.
[23] *NIST SP 800-33* (2001), p. 19.
[24] *NIST SP 800-33* (2001), p. 19.

that automatically scans a flash drive upon its insertion and prevents the execution of any programs or code contained on it. In the case of the artificial eye, risk mitigation might include wearing sunglasses that block light of a particular frequency or simply informing the device's human host about the vulnerability and providing instructions on how to best react if the device should accidentally encounter light of the relevant color and brightness and automatically reboot itself.

In some cases, risk mitigation may not be able to reduce the risk associated with a particular neuroprosthetic device to a level that the individuals or organizations involved consider acceptable; in those circumstances, the decision may be made to simply not use such a device.

A. Classifying vulnerabilities, threat sources, and threats

Researchers have proposed a number of ways of describing and categorizing vulnerabilities, threats, and risks relating to implantable medical devices (IMDs); many of these proposed classification schemes may also be relevant for advanced neuroprosthetic devices. Below we consider several of these classification schemes.

1. Vulnerability models

Hansen and Hansen propose a model for identifying and categorizing both the conditions that are required in order for a particular vulnerability to exist for IMDs and the effects that can occur as a result of the vulnerability being triggered or exploited.

- Classifying the preconditions for vulnerabilities. In Hansen and Hansen's model, the 'etiology' or unique set of circumstances in which a particular vulnerability can be exploited or triggered can be described with reference to three factors: the degree of **physical proximity** to the implanted device that a particular threat source must possess in order for the vulnerability to be triggered or exploited; particular **device activity** that the implanted device must be carrying out (typically, either 'sensing,' 'actuating,' 'information processing,' or 'communicating') in order for the vulnerability to exist; and any particular **state of the device's host** (such as a certain body position or posture, heart rate, state of conscious awareness, movement, or other activity) that must exist in order for the vulnerability to be triggered or exploited.[25]

- Classifying the effects of compromised vulnerabilities. In Hansen and Hansen's model, the impacts or 'pathogenesis' of vulnerabilities within IMDs that have been triggered or exploited can be described with reference to two factors: the **component affected** (which can be either

---

[25] See Hansen & Hansen (2010).

the implanted medical device, an external system that interacts with the implanted device, or the biological organism of the device's host) and the permanence of the impact (which might, for example, dissipate after a short time on its own, even without any intervention or attempt to address the impact; create a change in the implanted device, external system, or host's biological organism that will remain until it is proactively addressed and treated; or remain permanently without the possibility of being counteracted, treated, or removed).[26]

2. Threat source models

In discussing security for IMDs, Halperin et al. present the following types of adversaries, while acknowledging that other types may be possible:

- **Passive adversaries** are those that "eavesdrop on signals (both intentional and side-channel) transmitted by the IMD and by other entities communicating with the IMD."[27]

- **Active adversaries** are those which in addition to simply eavesdropping "can also interfere with legitimate communications and initiate malicious communications with IMDs and external equipment."[28]

- **Coordinated adversaries** are multiple adversaries who "coordinate their activities—for example, one adversary would be near a patient and another near a legitimate IMD programmer."[29]

- **Insiders** who may constitute a threat source can include "healthcare professionals, software developers, hardware engineers, and, in some cases, patients themselves"[30] who have the desire and ability to launch an intentional attack or who have the capability of unintentionally triggering a vulnerability and its effects.

Halperin et al. also note that adversaries can also be distinguished based on whether they use 'standard equipment' (such as a legitimate 'device programmer' unit that is produced by the manufacturer of an IMD and intended for use by authorized medical personnel in monitoring, updating, reconfiguring, or controlling the IMD) or 'custom equipment' (which may include homemade surveillance or control devices whose nature or capacities render the devices illegal).[31]

---

[26] See Hansen & Hansen (2010).

[27] See Halperin et al., "Security and privacy for implantable medical devices" (2008), p. 35.

[28] See Halperin et al. (2008), p. 35.

[29] See Halperin et al. (2008), p. 35.

[30] See Halperin et al. (2008), p. 35.

[31] See Halperin et al. (2008), p. 35.

3. Threat and attack models

a. Passive vs. active attacks

Cho and Lee classify potential threats as either passive or active attacks on implanted biosensors. In their framework:

- **Passive attacks** are those in which an adversary can "eavesdrop and collect the exchanged messages between a central device and each biosensor," although the adversary is unable to inject any payload into that central device and has "no ability to make a physical attack" on the implanted device; geolocating a device's host by eavesdropping on the device's transmissions would be a form of passive attack.[32]

- **Active attacks** on an implanted biosensor are those that incorporate "injecting/ modifying/blocking payload as well as eavesdropping;" they include denial of service attacks and spoofing attacks that may allow an adversary to impersonate other devices in their communication with the biosensor (or impersonate the biosensor in its communication with other devices).[33]

Cho and Lee suggest that the range of possible active attacks on an implanted biosensor does not include attacks involving direct physical access to the biosensor, as gaining physical access to such an implanted biosensor would be "impossible."[34] While such attacks might admittedly be complex and difficult (and, in many cases, undesirable for adversaries), we would argue that it is injudicious to exclude *a priori* the possibility of an active attack that does involve an adversary gaining physical access to an implanted device. It is undoubtedly true that many adversaries who have the desire and ability to remotely attack an individual's implant through wireless transmissions would have neither the desire nor ability to physically assault and mutilate the individual in order to gain direct physical access to his or her implanted device. However, the possibility cannot be ruled out; moreover, with future advances in nanotechnology and other technologies, it may someday become possible, for example, for an adversary to gain direct physical access to another person's implanted device not by performing a surgical operation but simply by spiking the person's beverage with an undetectable nanorobot cloud which, after ingestion, will travel through the person's body to the device.

b. Eavesdropping, impersonation, and jamming attacks

Ankarali et al. take a different approach in classifying attacks on implantable medical devices. They categorize malicious attacks on IMDs into three

---

[32] See Cho & Lee (2012), p. 208.
[33] See Cho & Lee (2012), p. 208.
[34] Cho & Lee (2012), p. 208.

categories: eavesdropping, impersonation, and jamming attacks. In their model:

- **Eavesdropping** attacks are those that "compromise the secrecy of transmitted private data e.g., patients' personal information, medical measurements, user location, and information that may be used to perform additional attacks [...]"[35]

- **Impersonation** is an attack in which an adversary impersonates an authorized user or system and is erroneously authenticated by an IMD. An adversary who has successfully impersonated an authorized user could potentially install malware on the IMD, steal information stored within it, cause it to behave in a way that would quickly exhaust its battery, or directly affect the host's organism in a way that could potentially be harmful or fatal.[36]

- **Jamming** is an attack in which an adversary undertakes a denial of service attack "by flooding the operating frequency of medical devices with an irrelevant signal," which can make it very difficult for a targeted IMD to carry out communication with external systems or other implanted devices that may be necessary for the IMD's successful functioning.[37]

c. Electronic, biological, psychological, or hybrid attacks

The threat models described above deal with attacks that are launched against neuroprosthetic devices in their nature as electronic devices. We would note, however, that such narrow, electronically focused models are inadequate to capture the full range of threats relating to neuroprosthetic devices. Because neuroprostheses are integrated into the neural circuitry of their human hosts, it is possible for an adversary to attack the host-device system by targeting either a neuroprosthetic device or the human being with whose structures and processes the device is so closely interconnected. A direct attack on the neuroprosthetic device itself might indeed typically target the device in its nature as an electronic computing device (e.g., through the use of computer viruses or worms or the use of a botnet to launch a denial of service attack). However, the possibility of neuroprosthetic devices that include biological material as key components[38] – or which even consist entirely

---

[35] See Ankarali et al., "A Comparative Review on the Wireless Implantable Medical Devices Privacy and Security" (2014), p. 247.

[36] See Ankarali et al. (2014), pp. 247-48.

[37] See Ankarali et al. (2014), p. 247.

[38] For the possibility of neuroprosthetic devices that include biological components, see Merkel et al. (2007). A hybrid biological-electronic interface device (or 'cultured probe') that includes a

of biological material – opens the door to attacks that target a neuroprosthetic device (or component thereof) in its nature as a biological system rather than an electronic device. Such attacks might, for example, utilize genetically engineered biological (rather than computer) viruses, radiation, chemical toxins, or other vectors or agents that can damage living biological organisms or disrupt their functioning. The human host of a neuroprosthetic
device could be subjected to a similar biological attack. Moreover, he or she
is also vulnerable to a different kind of psychological attack that operates not
primarily at the cellular level of electrochemical activity but at the higher
level of cognitive activity such as perceptions, beliefs, and volitions. We can
categorize such attacks in the following way:

- **Electronic attacks** are directed at a neuroprosthetic device in its nature
  as an electronic sensor, computing device, or effector. Such attacks
  may utilize means that are often employed against traditional computers, such as computer viruses or worms, other kinds of malware,
  or denial of service attacks.[39] While such attacks target the neuroprosthesis in its nature as an electronic device, the attack itself does
  not necessarily need to be delivered through electronic means; while
  a computer worm would typically be delivered through some electronic medium, an attack which (for example) uses a chemical agent
  or magnetic field to erase data stored on a neuroprosthetic device's
  internal magnetic storage unit would also be considered an electronic
  attack.

- **Biological attacks** target either a neuroprosthetic device or its human
  host in their nature as biological organisms. A biological attack could
  only be launched against a neuroprosthetic device that contains at
  least some biological material as components. Biological attacks can
  involve any agents or vectors that are capable of harming or disrupting the functioning of biological organisms, including bacteria, biological (rather than computer) viruses, chemical toxins, electromagnetic radiation, or excess heat, pressure, or other environmental conditions.

- **Psychological attacks** target the human host of a neuroprosthetic device at a level that overpowers, disrupts, or otherwise involves the

---

network of cultured neurons on a planar substrate is discussed by Rutten et al. in "Neural Networks on Chemically Patterned Electrode Arrays: Towards a Cultured Probe" (2007). Hybrid biological-electronic interface devices are also discussed in Stieglitz, "Restoration of Neurological
Functions by Neuroprosthetic Technologies: Future Prospects and Trends towards Micro-,
Nano-, and Biohybrid Systems" (2007).

[39] See Rao & Nayak (2014), pp. 141-62, for a discussion of such threats.

host's cognitive processes at a higher level than that of basic bio-chemical processes. Social engineering attacks[40] are a form of psycho-logical attack. Other such attacks might target a host's memories, emotions, or beliefs by providing particular kinds of sense data (e.g., comprising written text, audiovisual materials, or social interactions) whose harmful effect is generated not by its immediate impact on the host's sensory organs at the biochemical level but through the cogni-tive processes by which the host perceives and interprets the experi-ence and incorporates it into his or her memory. At present, psycho-logical attacks can only be launched against the human host of a neu-roprosthetic device; however, if a future neuroprosthetic device pos-sessed a sufficiently sophisticated form of artificial intelligence, it could potentially be subject to psychological attacks as well.[41]

• **Hybrid attacks** utilize more than one attack modality at the same time. An example might be a computer virus that infects an implanted neu-roprosthetic device, takes control of the device, and uses it to produce chemical agents that are released into the bloodstream of the device's human host in order to effect a biological attack against the host. Conversely, a genetically engineered virus or other biological agent could be introduced into a host's bloodstream, where it infects the biological component of an implanted neuroprosthetic device and in-troduces instructions in the form of genetic material (rather than ex-ecutable computer code) that will cause the biological component of the neuroprosthetic device to influence, disrupt, control, or even re-program the electronic component of the neuroprosthetic device in such a way that will constitute an electronic attack.[42]

The growing convergence of fields such as neuroscience, genetic engineer-ing, bionanotechnology, and molecular computing[43] along with the increas-ing integration of the human organism into biocybernetic and neurocyber-netic systems means that hybrid attacks against neuroprosthetic devices and their host-device systems are likely to become increasingly common. The

---

[40] See Rao & Nayak (2014), pp. 307-23.

[41] See Friedenberg, *Artificial Psychology: The Quest for What It Means to Be Human* (2008), for an in-depth analysis of psychological effects to which future AI systems (and in particular, arti-ficial general intelligences) might be subject.

[42] Some future neuroprosthetic devices with biological components might conceivably be de-signed to store program or data files in the form of genetic material; a virus or other biological agent that is able to rewrite the contents of that genetic material could, in effect, 'reprogram' the device – potentially in a way that would in turn lead to an alteration in the contents of program or data files stored by the device in electronic format. For a discussion of the possibilities of using DNA as a mechanism for the storage of data, see Church et al., "Next-generation digital infor-mation storage in DNA" (2012).

[43] See Chapter Three of this text for a discussion of this convergence.

most sophisticated kinds of attack agents may even be able to shift repeatedly between electronic, biological, and psychological forms while preserving intact their specified payload and moving toward their ultimate target. The existence of such threats will require information security practitioners to collaborate closely with experts from fields such as neuroscience, biology, biomedical engineering, epidemiology, cybernetics, and other disciplines, in order to develop full-spectrum information security plans, mechanisms, and practices that can protect against electronic, biological, psychological, and hybrid attacks.

d. Attacks on, through, or with a neuroprosthetic device

We can also distinguish three different kinds of attacks that relate to neuroprosthetic devices by identifying the role that a neuroprosthesis plays in the attack:

- **Attacks on a neuroprosthetic device** are targeted against a particular neuroprosthetic device or its host-device system. While they might involve an electronic vector such as a computer virus that requires the adversary who is launching the attack to possess significant knowledge of the targeted device, they might instead involve a purely physical attack (e.g., using radiation, electricity, chemical agents, or a physical weapon such as a hammer) that does not require the adversary to possess a neuroprosthesis or, in some cases, even to know exactly how to access or control the targeted device.

- **Attacks through a neuroprosthetic device** occur when an adversary hijacks (either temporarily or permanently) or otherwise exercises operative control over a neuroprosthetic device of which he or she is not the human host in order to utilize it in executing an attack. For example, if an individual possesses a cochlear implant that can record and wirelessly transmit live audio of everything that the person is hearing, an adversary might illicitly access that device and its live audio feed not in order to conduct surveillance on the person in whom the device is implanted but in order to eavesdrop on the conversations of two people who at the moment happen to be sitting in a train car near that person and who are the adversary's ultimate target.[44]

- **Attacks with a neuroprosthetic device** occur when an adversary is the human host of a neuroprosthetic device that he or she uses as a tool for planning or executing attacks. The potential use of neuroprosthetic devices by adversaries in this manner raises complex biomedical, legal, and ethical questions for information security practitioners, insofar as certain kinds of countermeasures that might be utilized to

---

[44] See Chapter Three of this text for further discussion of such possibilities.

prevent a cyberattack or mitigate an ongoing attack could potentially have a direct negative impact on the physical and psychological health of the adversary conducting the attack, if the computer that the adversary was using to manage the attack (and which is affected – and possibly disabled – by the countermeasures) is a neuroprosthesis that is integrated into the neural circuitry of the adversary's biological organism.[45]

Depending on its nature, it is possible for a single attack to be committed simultaneously on, through, and with neuroprosthetic devices.

e. Other notable types of attacks

Researchers have identified other kinds of possible attacks that technically can be classified using one of the schemes above (e.g., as either 'passive' or 'active'), but which merit specific mention here either because they have the potential to cause particularly devastating effects or are unique to neuroprosthetic devices. Such attacks include:

- **Long-term neural modifications.** Denning et al. note that some attacks on neuroprosthetic devices might be intended by an adversary to generate some long-term changes in the cognitive processes of a device's host.[46] In addition to affecting biological processes such as the functioning of internal organs, such changes could potentially impact the host's memories, beliefs, personality, skills, habits, relationships, and other activities and phenomena rooted in cognitive processes and abilities.

- **Disruption of neural networks.** By gaining illicit access to a neuroprosthetic device and disrupting or controlling its functioning, an adversary could potentially harm the device's host either by causing the death of existing cells or by disrupting neural networks through the growth of new cells or changes to existing cells.[47]

- **Induced addictions.** Some kinds of neuroprosthetic devices may be designed to function (or, if damaged or maliciously reprogrammed, be capable of functioning) in a way that could create an addiction on the part of a device's human host. Denning et al. note that some neuroprosthetic devices may reduce pain, change their host's mood, or activate pleasure and reward centers within the brain[48] in a way similar

---

[45] For a discussion of some of the legal and ethical issues that are involved, e.g., with the contemporary use of offensive countermeasures to mitigate botnets, see Leder et al., "Proactive Botnet Countermeasures: An Offensive Approach" (2009). The issue is also discussed further in Chapter Three of this text.

[46] See Denning et al. (2009).

[47] See Denning et al. (2009).

[48] See Denning et al. (2009).

to that found with opioid analgesic drugs. Such euphoric or other effects could potentially be produced either indirectly, by releasing chemicals or other agents into the host's bloodstream, or directly by electrically stimulating relevant portions of the host's brain.

We would note that an adversary who is able to take control of such a neuroprosthetic device could potentially induce an addiction on the part of its host (perhaps even gradually, without the host fully realizing what was occurring) and then threaten the host with termination of the addictive phenomena (and the infliction of resulting withdrawal symptoms) unless the host performs or fails to perform some action, as instructed by the adversary. The manufacturers or operators of certain kinds of neuroprosthetic devices could potentially use a similar approach conceptualized not as an 'attack' but as a 'business strategy' designed to increase the painful switching costs that a device's host would experience if he or she were to discontinue use of the device or to require ongoing purchases or subscriptions in order to continue experiencing the addictive neuroprosthetic effects.[49]

- **Resource depletion attacks.** Resource depletion attacks are a kind of denial of service attack that aims to disable or disrupt an implanted device by exhausting one of its critical resources – typically the power supply contained within its internal battery.[50] Resource depletion attacks are of critical concern for implantable neuroprosthetic devices, because after its implantation a device's store of certain resources (such as electrical power or chemical agents to be injected into the host's bloodstream) may be quite limited and not easily replenishable. Resource depletion attacks can easily be launched against implanted devices if their security controls have not been appropriately designed and implemented: for example, if an implanted device carries out some specified action that consumes electrical power (such as transmitting an RF signal to the external system in response) every time it detects that a wireless access request has been received from an external system, an adversary could potentially quickly exhaust the implant's battery and disable the implant simply by sending an unending series of wireless access requests to the device, even if all of the access requests were rejected by the implant.

---

[49] For an analysis of how such possibilities have been explored within a fictional setting, see Maj, "Rational Technotopia vs. Corporational Dystopia in 'Deus Ex: Human Revolution' Gameworld" (2015).

[50] See Hei & Du, "Biometric-based two-level secure access control for implantable medical devices during emergencies" (2011), and Freudenthal et al. (2007).

VI. Conclusion

Having considered fundamental information security concepts such as the CIA Triad, administrative, logical, and physical security controls, authentication and authorization, and the distinction between vulnerabilities, threats, and risks, in the next chapter we will explore in more detail many of the unique challenges that arise when attempting to apply these principles in the case of advanced neuroprosthetic devices.

Chapter Three

Critical Challenges in Information Security for Advanced Neuroprosthetics

Abstract. This text investigates unique challenges and opportunities that arise when one applies the general principles and practices of information security to the particular domain of advanced neuroprosthetics. Issues discussed include the distinction between a neuroprosthetic device and its host-device system; the need for a device to provide free access to outside parties during medical emergencies while rigorously restricting access to outside parties at other times; challenges relating to implanted devices' limited power supply and processing and storage capacities, physical inaccessibility, and reliance on wireless communication; new kinds of biometrics that can be utilized by neuroprostheses; complications and opportunities arising from the use of nontraditional computing structures and platforms such as biomolecular computing and nanorobotic swarms; and psychological, social, and ethical concerns relating to the agency, autonomy, and personal identity of human beings possessing advanced neuroprostheses. Having considered such issues, we discuss why traditional InfoSec concepts that are often applied to general-purpose computing and information systems are inadequate to address the realities of advanced neuroprosthetic host-device systems and why the creation of new specialized conceptual models of information security for advanced neuroprosthetics is urgently required.

I. Introduction

Information security is a large and complex field. While there are fundamental information security principles whose relevance is universal, the ways in which these principles are applied and elaborated in particular circumstances is subject to specialized practices and bodies of knowledge. The techniques used to secure a large organization's archive of decades of printed personnel files are different than those used to secure a factory's robotic manufacturing systems or an individual consumer's smartphone.

As with all kinds of information systems that have been developed by humankind, advanced neuroprosthetic devices present a unique array of information security problems and possibilities that exist within a particular set of technological, legal, political, ethical, social, and cultural contexts.[1] In this chapter we highlight a number of issues that may not be relevant for many other kinds of information systems but which give rise to considerations that are critical for the information security of advanced neuroprosthetic devices. Many of the issues discussed below constitute recurring themes that will be revisited in different contexts throughout the rest of this book.

II. Distinguishing a device user from a device host

In the case of a smartphone, the person who possesses the device and carries it with himself or herself on a daily basis is typically also the primary user and operator of the device: the smartphone's possessor powers it on and off, uses it to browse the web, check email, make calls, and play games, and downloads, installs, and uninstalls apps at will. In some institutional settings, a smartphone that is owned by the organization may not be controlled entirely by the person who possesses it; the organization's IT staff might remotely control and monitor some aspects of the phone's behavior and might, for example, restrict its possessor's ability to install new apps. However, the person possessing the phone still has a significant ability to control the device's settings and operation and to use its functionality to achieve his or her personal ends.

With advanced neuroprosthetic devices (and implantable medical devices generally) the situation can be quite different: a device's host – i.e., the human being in whose body the neuroprosthetic device is implanted and in which it operates – may have no ability whatsoever to control the device or to utilize its functionality for particular ends chosen by that person. Moreover, he or she may potentially not even realize that the device exists and has been integrated into his or her neural circuitry.

A. Shared operation and use by the host and an external party

In the case of an artificial eye, for example, it might be the case that the device's human host has full operational control over the device: through an act of volition, the host transmits instructions that cause the eye to focus its gaze on particular objects, and by using a standard computer the host can wirelessly connect to the artificial eye's internal computer, log onto a web-based command console, and perform remote diagnostics and software maintenance on the device.

---

[1] For an overview of ethical issues with ICT implants – many of which are relevant for advanced neuroprosthetics – see Hildebrandt & Anrig, "Ethical Implications of ICT Implants" (2012). For ethical issues in information security more generally, see Brey, "Ethical Aspects of Information Security and Privacy" (2007).

On the other hand, it might be the case that the device's human host is only able to control some limited aspects of the artificial eye's functionality (such as determining its focus or dilating its synthetic pupil) but has no ability to log into the device to manage diagnostic or maintenance tasks; a remote support team of specialized medical and biomedical engineering experts (e.g., working at the company that designed the device or the hospital that implanted it) may be the only party with such access – regularly monitoring the device's functioning, performing remote software upgrades and reconfiguration, and managing aspects such as the device's power consumption and synaptic stimulation strength. In such a scenario, the human being in whom the artificial eye was implanted would be the device's human *host,* but the host and the remote medical team would share a joint role as the device's *users* or *operators* who determine how and for what purposes it will be used and who controls its functionality.

B. Separate operator and host

It is also possible to envision circumstances in which a device's human host would play no role at all in controlling the device, determining the purposes for which it will (or will not) be employed, or managing its functionality. In such cases, another human being or organization may serve as the sole user and operator of the device, or, if the device possesses sufficiently sophisticated artificial intelligence, the device could even be said to be its own user and operator.

For example, a human host suffering from a particular illness may have been implanted with an endocrine neuroprosthetic device that can stimulate the host's thyroid gland and cause it to secrete hormones that affect the body's basal metabolic rate; however, the host has no means or access by which to control (or even directly influence) the functioning of the device, as it can only be controlled through a remote system that wirelessly transmits instructions to the device and is managed by an expert medical team. The medical team constitutes the device's sole user and operator, as team members decide when, whether, and to what extent the host's basal metabolic rate should be raised or lowered, and the medical personnel determine the objective (e.g., to facilitate weight loss or weight gain) that they are seeking to achieve through their management of the device.[2]

---

[2] Another possible cybernetic model is for a system to include an implanted component that gathers from its host's brain selected real-time data that is then transmitted to an external computer and used as input for a real-time computational model and simulation of the brain that allows the system's operators to determine what signals the implanted component should transmit to neurons in order to generate desired effects; the external computer then transmits those instructions to the implanted component, which stimulates neurons in accordance with those

C. Unwitting hosts

Especially in the case of noninvasive neuroprosthetic devices that can connect with the neural circuitry of a human being simply by touching the person's body (or even without touching it, through the use of wireless transmission and sensation), it may be possible that the human being with whose neurons a device is interacting may not even realize that the device exists or is in use. Even in the case of neuroprosthetic devices that can only function within the body of their human host, a device's host may potentially not realize that a neuroprosthetic device is located and operating within his or her body – e.g., if unbeknownst to the subject the device was implanted during the course of some other surgical procedure to which the host had consented, or if the device comprised components (such as a swarm of nanorobots) that could be unwittingly ingested simply by consuming a beverage.[3]

D. Physical and psychological damage caused to a user or host

In numerous contexts throughout this volume, we cite the possibility that a neuroprosthetic device that is poorly designed and operated, suffers a malfunction, or is compromised by a cyberattack could potentially subject its human host to significant physical or psychological harm – e.g., by impairing the functioning of the host's internal organs or providing a stream of sensory data that causes severe pain. While such harm to a device's host may be the more common hazard, with some kinds of neuroprosthetic devices it possible that such incidents might cause physical or psychological harm to the device's human *operator* rather than its *host*.

For example, imagine that a human host has – without his or her knowledge – been implanted with an advanced neuroprosthetic device located in the brain that detects sense data arriving from the optic and cochlear nerves and wirelessly transmits it to another human being – the device's operator – who through the use of a virtual reality headset and earphones is essentially able to see and hear all that the neuroprosthetic device's human host sees and hears. An adversary could potentially gain unauthorized access to the device's internal computer, reprogram it, disable its safety features, and alter its output so that rather than transmitting to the device's remote human operator a stream of the actual visual and auditory sense data received by the device's host it instead transmits signals which, in the operator's VR headset and earphones, produce an emission of blinding light and deafening noise that are powerful enough to both damage the operator's sensory organs, cause confusion and disorientation, and inflict major psychological distress.

---

instructions. See Lee et al., "Towards Real-Time Communication between in Vivo Neurophysiological Data Sources and Simulator-Based Brain Biomimetic Models" (2014).

[3] For the possibility that human hosts might unwittingly be implanted with RFID devices, see Gasson, "Human ICT Implants: From Restorative Application to Human Enhancement" (2012).

Note that if the operator (and, in this case, end user) of the neuroprosthetic device were receiving the transmitted sense data not through a conventional VR headset and set of earphones but through a neuroprosthetic implant within his or her own brain, then an arrangement would exist in which that individual was simultaneously serving as the host of one neuroprosthetic device and the operator of (potentially) two devices, and the damage would result from the person's roles both as host of his or her own implanted device and operator of the remotely controlled one.

III. Critical, noncritical, or no health impacts for a device's host and user

A conventional desktop computer is unlikely to have a direct critical impact (either positive or negative) on the health of its human user. While it is possible to imagine a critical health impact (e.g., if the computer electrocuted a user who had removed the outer casing and was attempting to repair the device, or if the user dropped the computer and injured himself or herself while attempting to carry it to a new location), such impacts involve highly unusual circumstances and do not result directly from the success or failure of the computer to perform its intended regular functions. It is perhaps more likely for a computer to have an impact on its user's health that is critical but highly indirect – e.g., by allowing its user to contact emergency medical personnel and summon assistance relating to some urgent health emergency or to research symptoms that the user was experiencing and diagnose a major illness.

Because of their intimate integration into the body and biological processes of their human host, however, neuroprosthetic devices have the potential to directly generate critical health impacts for human beings that other kinds of information systems are unlikely or unable to produce.[4] For example, the failure of a neuroprosthetic device that is responsible for regulating its host's respiratory or circulatory activity could result directly in the host's death within a matter of minutes or even seconds; conversely, the proper functioning of the device may extend its host's lifespan by many years. Such devices clearly possess a critical health impact.

Other kinds of technology, such as a neuroprosthetic robotic leg, may typically have a significant but indirect and noncritical impact on the health of its user. When functioning properly, the leg allows its user to stand and balance without falling and to walk, run, and exercise – all of which can contribute significantly (if not critically) to the user's health. On the other hand, if

---

[4] For some of the health impacts that can be generated, for example, by IMDs (whether neuroprosthetic devices or other kinds of IMDs), see Ankarali et al., "A Comparative Review on the Wireless Implantable Medical Devices Privacy and Security" (2014).

the device experiences a malfunction that gives the user a mild electric shock or requires the user to drag the immobile leg, such occurrences would constitute undeniably negative but likely noncritical health effects. Such device malfunctions call for prompt attention and maintenance but do not require an immediate response in order to save the life of the device's host.

Even with devices that generally demonstrate noncritical health impacts, it may be possible for them to yield a critical health impact in particular circumstances. For example, if an artificial eye were to fail while its host were driving an automobile or flying a helicopter, this could potentially have fatal consequences not only for the host but also for many other completely unrelated individuals. A similarly critical negative health impact could result if an adversary gained unauthorized access to the eye's internal computer and manipulated the device's sensory output to make its host believe, for example, that he or she were picking up a wooden stick from the lawn when in fact it was a poisonous snake.[5]

No matter how safe, limited, and benign its functionality might be, it is unlikely that any neuroprosthetic device could ever possess *no* health impact, insofar as it is by definition interacting with the neural circuitry of its human host. While some advanced devices might theoretically be able, for example, to remotely detect and interpret a human being's cognitive activity simply by relying on the passive capture of radiation or other phenomena naturally emitted by the person's brain – and thus allow the person to control some remote robotic system through thought alone – such a device would not be considered a 'neuroprosthetic device' according to the definition of this text, since it is not truly *integrated into* the neural circuitry of the person's brain or body in any substantive sense.

Note that when a neuroprosthetic device possesses (or has the potential to demonstrate) a critical health impact, ensuring information security for the device and its host and user becomes even more important than usual. The greatest possible attention to information security must be shown by such a device's designer, manufacturer, operator, and host; such a device is likely to need extremely robust and stringent security controls that may not be necessary for devices with a lesser potential health impact. This obligation to ensure the highest possible degree of information security for neuroprostheses with a (potentially) critical health impact will inform our discussions of many specific security practices and mechanisms throughout this text.

---

[5] For the possibility that sensory neuroprostheses might be used to supply false data or information to their hosts or users, see McGee, "Bioelectronics and Implanted Devices" (2008), p. 221.

IV. Lack of direct physical access to implanted systems

Because of the risks and difficulties involved with surgical procedures, after a neuroprosthetic device has been implanted in its human host, information security personnel working to protect that device and its host may never again enjoy the opportunity to physically inspect, manipulate, or otherwise access the device; from that point forward, the only means of interacting with the device may be through wireless communication[6] (assuming that the device possesses such capabilities) or through action of the device's host (e.g., if the host has the ability to communicate with, influence, or control the device through acts of volition or other internal cognitive or biological processes). On the one hand, such limitations in physical access may create challenges for InfoSec personnel: for example, it may be necessary to build into a device at the time of its creation and implantation security controls that will be powerful and adaptable enough to counteract threats that may not yet even exist and will be developed only years or decades after the device's implantation. Moreover, some physical security controls that can be applied to conventional computers (such as the use of hardwired rather than wireless communication between devices) may be impossible to apply to an implanted device that exists beyond one's grasp and direct physical control. On the other hand, the fact that a neuroprosthetic device has been implanted in a human host might also bring some security benefits (such as physical concealment of the device's existence[7]) that are more difficult to implement in other kinds of information systems such as desktop computers.

V. Requirements for 100% availability

From the perspective of information security, the information contained within particular information systems can be said to demonstrate availability if the "systems work promptly and service is not denied to authorized users;"[8] other information security experts have defined availability as "Ensuring

---

[6] For an overview of information security issues relating to the wireless communication of IMDs such as body sensor networks (BSNs) – many of which are relevant for advanced neuroprosthetics – see Ameen et al., "Security and Privacy Issues in Wireless Sensor Networks for Healthcare Applications" (2010).

[7] Regarding the role of physical concealment in the protection of information systems, see *NIST Special Publication 800-53, Revision 4: Security and Privacy Controls for Federal Information Systems and Organizations* (2013), pp. F–205-06.

[8] *NIST Special Publication 800-33: Underlying Technical Models for Information Technology Security* (2001), p. 2.

timely and reliable access to and use of information"[9] or the "Usability of information for a purpose."[10] These are the meanings of availability intended, for example, when the word is used as part of the CIA Triad of InfoSec objectives.

Other branches of computer science and information technology, however, use the word 'availability' with an equally specific but somewhat different meaning: in that alternative sense, an information system's availability is a quantitative measure of how likely it is that the system will be operational (i.e., not out of service due to some hardware or software fault) at a given point in time. One can quantify a computer's **reliability** as the **mean time to failure** (MTTF), or the average length of time that a system will remain in continuous operation before experiencing its next failure,[11] while the **mean time to repair** (MTTR) is the average length of time needed to detect and repair a failure after it has occurred and thereby return the system to operation. A computer's steady-state **availability** A can be defined as the likelihood that the system is operating at a particular moment; it is related to the system's MTTF and MTTR by the equation:[12]

$$A = \frac{\text{MTTF}}{\text{MTTF} + \text{MTTR}}$$

A typical requirement for general-purpose commercial computer systems is that they demonstrate 99.99% availability over the course of a year.[13] However, in the case of some neuroprosthetic devices, that level of availability could be wholly unacceptable, insofar as it would represent an average of roughly 53 minutes of downtime over the course of a year. With some kinds of advanced neuroprosthetic devices that regulate critical circulatory or respiratory functions within their host's body, the impact of a device ceasing to operate for a period of 53 consecutive minutes could prove fatal to its human host. On the other hand, for other kinds of devices, a period of 53 consecutive minutes in which the device was nonfunctional might not be particularly harmful – especially if the outage took place as part of a scheduled repair

---

[9] 44 U.S.C., Sec. 3542, cited in *NIST Special Publication 800-37, Revision 1: Guide for Applying the Risk Management Framework to Federal Information Systems: A Security Life Cycle Approach* (2010), p. B-2.

[10] See Parker, "Toward a New Framework for Information Security" (2002), p. 124.

[11] See Grottke et al., "Ten fallacies of availability and reliability analysis" (2008), as discussed in Gladden, "A Fractal Measure for Comparing the Work Effort of Human and Artificial Agents Performing Management Functions" (2014), from which this section on availability is adapted.

[12] See Grottke et al. (2008).

[13] See Gunther, "Time—the zeroth performance metric" (2005).

process and a medical support team was ready to monitor and treat the device's host during that period, or if the host possessed a reliable backup system that he or she could easily activate during the one hour a year when the primary device typically became nonfunctional.

For some kinds of neuroprosthetic devices, a single long outage once per year might be less harmful than many frequent outages of shorter duration. For example, a system that freezes up and becomes nonfunctional for one millisecond out of every 10 seconds would also demonstrate roughly 99.99% availability. If such a neuroprosthetic device serves as the controller of a complex device network in which it receives data from and coordinates the actions of a number of other implanted devices regulating its host's core biological functions, such frequent outages could potentially impair the work of the entire system – especially if the system needs a couple of seconds to confirm the system's integrity and regain full operational capacity after each millisecond outage.

The question of how much time is needed to *fully recover* complete operational capacity by a system whose outage has already ended and which is already nominally functional raises another question relating to availability. In the sense just described here, availability has traditionally been understood in a binary manner: a system is either 'up' or 'down,' with no possible intermediate states. While the binary definition of availability is conceptually elegant and results in an equation that is easy to apply, it completely fails to capture many of the difficult realities with which IT professionals must often grapple. For example, if a hardware failure, software configuration error, or denial of service attack dramatically impacts an organization's information system and reduces its performance to only 0.5% of its normal processing speed and capacity, the lived experience of many people in the organization may be that the system is experiencing a catastrophic outage and is wholly nonfunctional. However, according to the technical definition of availability just given, one would say that the system has not 'failed,' because it has not failed *completely*; although the system's performance has been dramatically degraded, the system is still operational and functional – simply at a reduced level of speed and capacity. In order to address such limitations with the binary definition of availability, Rossebeø et al. argue that a more sophisticated measure for availability is needed that takes into account qualitative aspects of a system's performance and which recognizes a range of intermediate qualitative states between simply 'up' and 'down.'[14] This is especially true for many kinds of advanced neuroprosthetic devices.

---

[14] See Rossebeø et al., "A conceptual model for service availability" (2006).

For example, imagine that a high-resolution artificial eye provides its host with output sense data corresponding to a field of vision consisting of five million pixels; that is arguably roughly comparable to the resolution offered by a natural human eye.[15] When functioning at such a level, one could say that the system is fully operational and that both the information system (constituted by the artificial eye) and the system's information (constituted by raw sense data from the external environment that has been processed by the device and outputted to its human host) are available. However, during moments of especially high demand on the device's internal processor (e.g., when it is performing a major diagnostic operation or undergoing a software upgrade), the field of vision provided by the device's output data degrades to perhaps, say, only 5,000 total pixels. Although this would represent a dramatic 99.9% reduction in performance (as measured by resolution), it would still be sufficient to allow the device's host to carry out such tasks as navigating around a room, recognizing faces, or reading text on a computer screen.[16] If one were restricted to a binary definition of availability, one would need to say that the information system and its information were both 'available,' because the system was indeed functioning and making information available, if only to a limited degree.

Imagine further that a major sensor component within the artificial eye fails and the device switches instantaneously to a rudimentary backup sensor of quite limited capacity: as a result, the output data produced by the device represents a visual field of only 64 total pixels. Such limited visual data would likely be insufficient to give the device's host the ability to perform even basic tasks such as navigating visually around a room or recognizing a particular face. In this situation, a binary measure of availability would tell us that the information system and information are still available: the device, after all, is functioning and providing its human host a constant stream of data that represents (in a very limited fashion) the raw sense data received from the external environment. However, from the perspective of information security it would be difficult to say without qualification that the information system and its information were 'available' in the sense envisioned by the CIA Triad. If an adversary had launched a cyberattack and gained unauthorized access to the artificial eye in an effort to steal visual information that reveals where the device's host is and what he or she is doing, the adversary would likely

---

[15] The question of the human eye's 'resolution' is quite complicated. While an eye contains many more than one million sensors (i.e., individual rods and cones), there are only about one million output neurons (in the form of ganglion cells) that transmit data from the eye to the brain, and roughly half of the data comes from the tiny fovea, or focal point at the center of the field of vision in which the eye provides sharp central vision. See Linsenmeier, "Retinal Bioengineering" (2005), for a discussion of some such issues.

[16] See Weiland et al., "Retinal Prosthesis" (2005), and Viola & Patrinos, "A Neuroprosthesis for Restoring Sight" (2007).

not feel satisfied with gaining access to a display of 64 pixels that contains no practically interpretable, *useful* information about the host's external environment or activities.

When considering 'availability' in its sense of the functionality or operationality of an information system, one must thus carefully consider whether the measure should be defined in a binary manner in which 'availability' signifies any non-zero level of functionality; whether the measure should be defined in a binary manner in which 'availability' signifies a level of functionality greater than some particular specified non-zero threshold; or whether a non-binary, multivalent understanding of availability is more appropriate. For certain kinds of neuroprostheses with critical health impacts, simply setting and meeting a target like 99.99% or even 99.999% availability (the so-called 'five nines' level of availability, equivalent to roughly five minutes of downtime per year) may not be sufficient, as such periods of downtime could be harmful or even fatal to the devices' human host. For some kinds of devices, the only target that is acceptable from a legal and ethical perspective may be that of 100% availability – even if it is known in advance that this may be unattainable, due to circumstances beyond the control of a device's designer, manufacturer, operator, or host.[17]

VI. The need for rigorous security vs. the need for instant emergency access

The fact that advanced neuroprosthetic devices are integrated into the neural circuitry of their human host creates a unique information security challenge – and potentially a dilemma – for the developers of such technologies.

A. The need to protect a device from unauthorized parties

On the one hand, a neuroprosthetic device ought to be better secured against computer viruses and unauthorized access than, say, its host's laptop computer or smartphone. After all, if a person's smartphone is compromised, it could potentially result in inconvenience, financial loss, identity theft, the disclosure of sensitive and embarrassing information, and potentially even legal liability for that person – but it is unlikely to have a direct critical impact

---

[17] The (presumably unachievable) goal of 100% availability is related to the concept of a 'zero-day vulnerability' in information security, in which software or hardware developers must work as quickly as possible to develop a patch or fix for an uncorrected flaw. While it is known that some time will be required to repair the flaw, InfoSec practitioners must work to do so as quickly as possible; the longer the delay in patching the flaw, the more likely that the vulnerability will be exploited and cause harm to device hosts and users.

on the person's physical health. If a person's neuroprosthetic device is compromised, though, this might not only allow an adversary to steal sensitive medical data and information about the person's cognitive activity (potentially even including the contents of sensory experiences, memories, volitions, fears, or dreams); it might also – through the device's impact on the person's natural neural networks – have the effect of rendering damaged, untrustworthy, or inaccessible the information stored within the host's natural biological systems, including the brain's memory-storage mechanisms.[18] The potential impact of the intentional manipulation or accidental corruption of a neuroprosthetic device could be potentially catastrophic for the physical and psychological health of its human host; this suggests that access to such a device should be limited to the smallest possible number of human agents who are critical to its successful functioning, such as the device's human host and – if a different person – its primary human operator.

B. The need to grant emergency device access to outside parties

On the other hand, though, there are reasons why restricting access to a neuroprosthesis too severely might also result in significant harm to the device's host. For example, imagine that the human host of a neuroprosthetic device has been involved in a serious accident or is unexpectedly experiencing an acute and life-threatening medical incident. In this case, emergency medical personnel on the scene may need to gain immediate access to the neuroprosthetic device and exercise unfettered control over its functionality in order to save the life of its host.[19] The same mechanisms that make it difficult for a cybercriminal to break into the neuroprosthetic device – such as proprietary security software, file encryption, and a lack of physically accessible I/O ports – would also make it difficult or impossible for emergency medical personnel to break into the device. In principle, government regulators could require (or the manufacturers of neuroprosthetic devices could voluntarily institute) mechanisms that allow such devices to be accessed by individuals presenting certain credentials that identify them as certified emergency medical personnel who are trained in the use of such technologies,[20] or neuro-

---

[18] For the possibility that an adversary might use a compromised neuroprosthetic device in order to alter, disrupt, or manipulate the memories of its host, see Denning et al., "Neurosecurity: Security and Privacy for Neural Devices" (2009).

[19] See Clark & Fu, "Recent Results in Computer Security for Medical Devices" (2012); Rotter & Gasson, "Implantable Medical Devices: Privacy and Security Concerns" (2012); and Halperin et al., "Security and privacy for implantable medical devices" (2008) – all of whom who make this point regarding IMDs. Halperin et al., especially, consider this question in detail.

[20] For a discussion of such certificate schemes and related topics, see, for example, Cho & Lee, "Biometric Based Secure Communications without Pre-Deployed Key for Biosensor Implanted in

prosthetic devices could be designed to temporarily disable some of their security controls if they detect that their host is experiencing a medical emergency.[21] However, such mechanisms themselves create security vulnerabilities that could potentially be exploited by an adversary who is highly motivated to gain access to the information contained in a neuroprosthetic device or its human host.

C. Proposed methods for balancing these competing demands

Ideally, a neuroprosthetic device (and especially one with a critical health impact for its host) would be utterly impervious to all attacks and impenetrable to any adversaries attempting to gain unauthorized access – but would instantaneously grant full access and place itself at the disposal of any trained and well-intentioned individual who, in time of need, was attempting to use the device to save its human host's life. In reality, it is difficult to design a device that simultaneously fulfills both of these visions, and trade-offs need to be made. The priorities that a particular host adopts for his or her neuroprosthetic device's information security plan may partly depend on what the host considers to be more likely: that that a corporate espionage agent or cybercriminal will someday attempt to break into his or her neuroprosthesis in order to steal the person's memories or arrange the person's death, or that an emergency medical technician will someday need to break into the device and override its programmed functioning in order to deliver life-saving medical treatment or prevent the host from suffering some grave neurological damage.

Hansen and Hansen note that the controls and countermeasures designed to protect implantable medical devices from unauthorized access while simultaneously ensuring that authorized parties (such as emergency medical personnel) receive access typically take one of three forms, as either **detective**, **protective**, or **corrective** countermeasures. For example, a security control that alerts a device's human host to a series of unsuccessful logon attempts would be detective, one that blocks wireless transmissions from reaching an implanted device would be protective, and one that allows compromised data within the device's internal memory to be replaced by backup data from an external system would be corrective.[22] Below we consider a number of specific controls and countermeasures that have been proposed to address the challenge of providing both rigorous protection for implanted devices and robust

---

Body Sensor Networks" (2012), and Freudenthal et al., "Practical techniques for limiting disclosure of RF-equipped medical devices" (2007).

[21] Regarding the ability of IMDs to detect a medical emergency that is being experienced by their human host, see Denning et al., "Patients, pacemakers, and implantable defibrillators: Human values and security for wireless implantable medical devices" (2010), pp. 921-22.

[22] See Hansen & Hansen, "A Taxonomy of Vulnerabilities in Implantable Medical Devices" (2010).

emergency access for authorized personnel. While some of these approaches may not be relevant for all kinds of neuroprostheses (especially devices that operate outside of their host's body), many of the approaches are relevant for implantable neuroprosthetic devices, and they offer an excellent starting point for considering issues of emergency access for neuroprosthetic devices more broadly.

1. Certificate schemes and predeployed keys managed by device manufacturers or operators

One approach to addressing this challenge involves the creation of a centrally managed worldwide certification scheme that would be administered either by a device's manufacturer or operator. At the time of its manufacture or implantation, a pre-configured backdoor key can be installed in the operating system of an IMD; the backdoor key can be maintained by the device's manufacturer or operator in a cloud-based system that can be accessed globally through the Internet by medical personnel who are treating the device's host during the course of a medical emergency.[23]

However, this model demonstrates significant limitations and disadvantages. Hei and Du note that this 'certificate' approach would fail in cases where the medical emergency (and treatment) were occurring in a location in which the personnel providing medical treatment did not have immediate Internet access (e.g., if an accident occurred in a remote wilderness area where wireless Internet access was absent or unreliable); moreover, they note that maintaining such a system of backdoor keys that are always accessible online would be complex and costly.[24]

2. An external hardware token whose possession grants access to an implanted device

Bergmasco et al. explore the use of hardware tokens such as a small USB token that can be inserted into a standard USB port as means of authenticating users of medical devices and information systems.[25] External hardware tokens that utilize wireless technologies – such as an RFID tag worn or carried by the host of a neuroprosthetic device – could potentially be used by the implanted device to wirelessly authenticate its user; even technologies such as a USB token that require physical insertion of the token into the implanted

---

[23] For a discussion of such matters, see, for example, Cho & Lee (2012) and Freudenthal et al. (2007).

[24] See Hei & Du, "Biometric-based two-level secure access control for implantable medical devices during emergencies" (2011).

[25] See Bergamasco et al., "Medical data protection with a new generation of hardware authentication tokens" (2001). That text does not specifically consider the case of implantable medical devices but instead considers access to medical devices and information systems more generally.

device could conceivably be used for authentication if the implanted device possesses a port, slot, or other component that is accessible from the exterior of its host's body. Denning et al. describe a similar approach of 'proximity bootstrapping' in which emergency medical personnel who may come into contact with patients possessing implantable medical devices could be given a portable unit that can communicate with an IMD through sound waves or 'physiological keying' when brought into contact with the device's host to request access to the IMD.[26]

Hei and Du note that such an external hardware token can be lost (thus denying emergency access to an authorized party who has misplaced his or her token) or stolen (thus potentially granting access to an unauthorized party who has stolen or found the device).[27] We would also note that unless there is international regulation or extensive industry-wide (and even inter-industry) collaboration between the manufacturers of diverse kinds of implantable devices to agree on a single shared scheme for such tokens, the need for emergency personnel around the world to carry a bewildering array of tokens for different manufacturers' devices (and perhaps to test them all on every newly encountered patient in order to check whether any implanted devices might exist) could become unwieldy. On the other hand, if all device manufacturers were to utilize a single shared token system, this would give potential adversaries a lone attractive target on which to concentrate all of their efforts to compromise such devices.

3. A cryptographic key stored on a host's person

One design for a security control is for wireless access to an implanted device to be secured using a cryptographic key (e.g., consisting of a string of characters) that must be possessed by any other implant or external device that wishes to access the implant. The cryptographic key – along with instructions to emergency medical personnel describing the nature of a host's implanted device and how medical personnel should access and configure the device during particular kinds of medical situations – could then be displayed on a standard medical bracelet worn by the device's host, similar to the sort that is already commonly used by individuals to alert emergency medical personnel to the fact that, for example, they suffer from diabetes or asthma or possess a pacemaker. However, such bracelets are not extremely secure: they can potentially be lost or stolen, may actually alert adversaries to the presence of an implantable device that they otherwise would not have known about,

---

[26] See Denning et al. (2010), p. 922.

[27] See Hei & Du (2011).

and (depending on their design) could potentially allow an adversary to photograph or otherwise obtain the information contained on the bracelet without even directly contacting it.[28] A card displaying the cryptographic key or password for an implanted device that is carried in the wallet of the device's human host[29] is less visible to potential adversaries while still likely to be found by emergency medical personnel treating the host.

Denning et al. and Schechter propose an access control method for implantable medical devices that utilizes ultraviolet-ink tattoos.[30] In this approach, an IMD is secured using a cryptographic key consisting of a string of characters. These characters can then be tattooed on the body of the device's host using a special ink that only becomes visible under ultraviolet light. Schechter notes that unlike a wearable accessory such as a bracelet that is used to store the cryptographic key, a tattoo cannot be lost or misplaced by a device's host; moreover, the existence of the tattoo is not readily apparent to potential adversaries, and if necessary the device's host could even prevent a suspected adversary from illuminating and reading the tattoo simply by applying sunscreen that sufficiently blocks ultraviolet light. In comparison to the use of a bracelet, one disadvantage of this approach is the fact that emergency medical personnel would have no immediate indication that the tattoo exists; they would only know to search for a tattoo if they had some particular reason for suspecting that the human subject whom they were treating might possess an implantable device secured by such a cryptographic key. Moreover, emergency medical personnel might not always have the correct sort of UV light available. On the other hand, if the use of UV-ink tattoos for such purposes someday became widespread, then it could conceivably become a standard practice for medical personnel to carry such UV lights and check all patients for the presence of such tattoos. An alternative would be a traditional tattoo that is visible to the naked eye,[31] which would be more likely to be noticed by emergency personnel but would also be more likely to alert an adversary to the existence of an implanted device and allow him or her to obtain the cryptographic key from the tattoo (e.g., since depending on its location on the host's body it could potentially be photographed from a distance).

Denning et al. note that the same sort of severe accident or injury that might require a device's host to receive emergency medical treatment might also damage or destroy the information contained in a tattoo or a medical

---

[28] See Hansen & Hansen (2010) and Schechter, "Security that is Meant to be Skin Deep: Using Ultraviolet Micropigmentation to Store Emergency-Access Keys for Implantable Medical Devices" (2010).

[29] See Denning et al. (2010).

[30] See Denning et al. (2010) and Schechter (2010).

[31] See Hansen & Hansen (2010).

bracelet worn on the host's person, thus creating a significant disadvantage for such approaches.[32]

4. Access control based on ultrasonic proximity verification

Rasmussen et al. propose a model of access control for implantable medical devices in emergency situations that relies on ultrasound technology to verify the physical proximity of an external system attempting to gain access to an IMD. Under normal circumstances, the IMD would require an external system to possess a shared cryptographic key in order to grant the external system access to the IMD; however if the IMD detects that its host is undergoing a medical emergency, it then shifts into an 'emergency mode' in which any external system is allowed to access the IMD, as long as it is within a certain predefined distance – with the distance gauged by measuring the time required for ultrasound communications to travel between the IMD and external system.[33]

Hei and Du note that while relying on such ultrasound proximity detection as a primary security control would be inappropriate in normal everyday circumstances (as the control would only function successfully if it could be assumed that the device's host would typically recognize an approaching adversary and prevent him or her from getting to close to the host), it could be appropriate for use in emergency circumstances; they also note that it could be difficult to integrate a sufficiently powerful and effective ultrasound receiver into some kinds of implantable devices.[34]

5. Physical radio frequency shielding of an implanted device

Hansen and Hansen suggest that a simple means of securing IMDs against wireless RF attacks would be for a device's host to wear a shielded undershirt or shielded bandages applied to the skin that block or disrupt wireless communications. They note that such electromagnetic shielding would be relatively lightweight (thus not greatly inconveniencing the device's host) and could easily be removed by emergency medical personnel who need to treat the host during a critical health situation.[35]

Such an approach is not without disadvantages when applied to advanced neuroprosthetic devices. For many human hosts, being required to wear a special shielded undershirt or bandages wherever they go (e.g., while at the

[32] See Denning et al. (2010), p. 920.
[33] See Rasmussen et al., "Proximity-based access control for implantable medical devices" (2009). Regarding the possibility of IMDs being able to detect a medical emergency that is being experienced by their human host, see Denning et al., "Patients, pacemakers, and implantable defibrillators: Human values and security for wireless implantable medical devices" (2010), pp. 921-22.
[34] See Hei & Du (2011).
[35] See Hansen & Hansen (2010).

beach or taking a shower) may be an undesirable inconvenience in their daily life, and some hosts might choose to temporarily remove the shielding during such situations, leaving them vulnerable to attack. Moreover, those hosts whose neuroprosthetic devices are implanted in their brain would need to wear a shielded hat, wig, bandage, or other appliance on their heads at all times, which can be more awkward and inconvenient than wearing a shielded undershirt and may also serve to draw adversaries' attention to the fact that the person possesses a cranial neuroprosthetic device.[36] Finally, such shielding would potentially block not only wireless RF attacks created by adversaries but *all* wireless RF communications; this would be impractical and even dangerous to the host's health and well-being if the implanted device requires periodic instructions communicated wirelessly from an external system in order to operator correctly (or if an external system needs to receive periodic communications from the implanted device – e.g., containing real-time medical data from the host – in order to correctly configure and apply medical treatments and ensure the host's safety and health). Indeed, in such cases, it would be imperative for a device's host to ensure that an adversary did *not* surreptitiously alter the host's clothing or provide the host with clothing that includes shielding that would disrupt the proper functioning of the host's neuroprosthetic devices.

6. Subcutaneous buttons that grant access to a device

Hansen and Hansen also suggest the possibility of a subcutaneous button that is implanted beneath the surface of a host's skin and which can be activated by pressing the host's body at a particular location. When the button is pressed, the IMD temporarily enters a special programming mode that disables some of its security controls and allows the device to be remotely accessed and reprogrammed (e.g., through wireless transmissions) for a specified period of time.[37] Hansen and Hansen note that such a button might be prone to being pressed accidentally, thus its location and nature would need to be carefully chosen in order to minimize such possibilities.

We would argue that while perhaps not appropriate as a sole security control, such a mechanism might be more effectively used in conjunction with other security controls. For example, if one needed to both press the button

---

[36] The 'tin foil hat' referenced within popular culture as a stereotypical tool used by paranoid individuals to protect themselves from telepathic attacks and mind control might thus – while not the most effective approach to neural defense – be an idea not entirely lacking in substance. For an analysis of the RF-shielding properties of such devices, see Rahimi et al., "On the effectiveness of aluminium foil helmets: An empirical study" (2005). For a discussion of 'psychotronics' in popular culture (as well as of supposed efforts on the part of military agencies to develop technologies that could potentially be used for remote mind control), see Weinberger, "Mind Games" (2007).

[37] See Hansen & Hansen (2010).

and possess a particular hardware token or remove shielding from the body in order to wirelessly access the implanted device, this would create greater security. Even including two different subcutaneous buttons in different parts of the body that need to be pressed simultaneously or within a certain window of time would increase security and reduce the likelihood of the device's wireless access controls being inadvertently disabled.

Another design question to be considered is whether the existence and nature of a subcutaneous button should be visible to the naked eye, visible only with the aid of particular equipment (such as an ultraviolet lamp), or wholly undetectable to outside parties. Making a button less easily detectable would decrease the chances that an adversary would discover its existence, while perhaps also making it more difficult for emergency medical personnel to notice that the host possesses an implanted neuroprosthetic device and successfully access it.

7. An external unit maintaining the secured state of an implanted device

Denning et al. propose a model in which the host of an implanted medical device carries a secondary, external device (which they call a 'Cloaker') that controls the access that other external systems can gain to the implanted device. While they propose and consider several different variations on that theme, all of the cloaking approaches of this type share in common the fact that when the cloaking device is present, the implanted device can only be accessed by certain authorized parties; when the cloaking device is absent (or nonfunctional), the implanted device 'fails open' into a state in which the implant responds to all access requests received from external systems. In the case of a medical emergency, medical personnel could access the implanted device simply by removing the cloaking device from the primary device's host.[38]

The IMD could potentially determine whether or not the external cloaking device is present by sending an "Are you there?" query to the cloaking device every time that some external system attempts to access the IMD; however, Denning et al. note that this could expose the IMD to a denial of service attack in the form of a resource depletion attack that attempts to exhaust the IMD's battery simply by sending an unceasing series of access requests. Denning et al. thus suggest an alternative approach in which the IMD sends an "Are you there?" query to the cloaking device at periodic, nondeterministic intervals set by the device's designer; the designer could choose intervals that are brief enough to ensure that the IMD will fail open quickly enough in the case of a medical emergency but not so brief that the IMD's battery will be exhausted

---

[38] See Denning et al., "Absence Makes the Heart Grow Fonder: New Directions for Implantable Medical Device Security" (2008).

through frequent queries. Denning et al. note that one challenge for this model is to deal effectively with the possibility that an adversary could jam the wireless communications between the IMD and cloaking device, thereby inducing the IMD to fail open into a nonsecure state in which the adversary can gain access to the implant. Another disadvantage is the fact that while the IMD's periodic "Are you there?" queries to the cloaking device may not consume large amounts of power, they do place some drain on the unit's power supply, which – as for many implantable devices – may be quite limited and difficult to replenish.

8. An external gateway that jams communication with an implanted device

Zheng et al. propose a model in which the host of an implanted device also wears or carries an external 'gateway' device that controls access to the implant. The gateway device not only jams wireless communication and blocks transmissions (e.g., from an adversary) from reaching the implanted device, but it also impersonates the implanted device and communicates with an adversary's system whenever it detects an adversary attempting to access the implant. Because all of the adversary's access requests are being received and processed by the external gateway rather than the implant, it is not possible for the adversary to subject the implant to a resource depletion attack and exhaust its battery or otherwise disrupt its functioning by flooding the implant with an unending series of access requests. In the case of a medical emergency, medical personnel who are treating the device's host need only locate and power off the external gateway device worn or carried by the host; as soon as the gateway has been disabled and its jamming and spoofing activities have ceased, direct wireless access to the implanted will be possible.[39] In a sense, this model is similar to the 'Cloaker' approach proposed by Denning et al.; however, it places no drain on the IMD's battery, since the IMD does not need to send periodic "Are you there?" queries (or otherwise transmit data to) the external component. It also eliminates the possibility that an adversary could impersonate the external cloaking device and send wireless signals to the IMD that force it to remain secured and inaccessible when the device's host is indeed undergoing a health emergency and medical personnel have removed the 'real' cloaking device from the host's person.

9. Audible alerts to increase a host's awareness of potential attacks

Halperin et al. note that some IMDs generate an audible alert that their host can hear when the device's battery is nearly exhausted, and they recommend that similar audible alerts be used as a supplemental security measure

---

[39] See Zheng et al., "A Non-key based security scheme supporting emergency treatment of wireless implants" (2014).

for IMDs: if a device detects suspicious activity that may indicate an attack (such as a series of unsuccessful attempts by a wireless user to access the device), the device could generate an audible alert that its human host would hear.[40] Halperin et al. note that while such an alert would not in itself directly block an ongoing attack against the device, the fact that the device's host has been alerted to the possibility of an ongoing attack means that the host could then take specific actions (e.g., as previously instructed by the device's operator) that would directly prevent or block an attack that was in progress.

However, Hei and Du note that an audible alert might not be noticed by the device's host if he or she were in a noisy environment (nor, we might add, if the host were asleep); they also note that a mechanism for generating an audible alert consumes electrical power and thus cannot easily be incorporated directly into an implantable device itself, insofar as power is a limited and very precious commodity for many such devices.[41] Halperin et al. propose to avoid such power constraints by implanting a secondary device whose sole purpose is to audibly alert its human host to attacks on the primary implanted device. They have developed an implantable prototype based on the Wireless Identification and Sensing Platform (WISP) that can harvest energy from an external power source in the form of a radio signal generated by a standard UHF RFID reader; in this way, the secondary WISP device places no demand on the power supply of the primary device implanted in the host.[42] The WISP device uses a piezoelectric element to generate an audible beep if it detects certain kinds of RF activity (such as a series of wireless access requests from an external RFID reader) that could indicate an attempt by an adversary to access the primary implanted device.

An alternative approach proposed by Halperin et al. similarly relies on the use of sound to make a device's host aware of a potential attack: they have developed a prototype implantable device that exchanges its symmetric cryptographic key with an external system using sound waves that are audible to the device's host and detectable by an external system in close proximity to the host's body but not detectible (e.g., to adversaries) at significant distances from the host's body.[43] In this way, whenever the device's host hears the relevant kind of sound generated by the implanted device, he or she knows that some external system has just submitted an access request to the implanted device and is receiving the cryptographic key. If that attempt is unauthorized, the host could potentially thwart it by moving away from that location and away from whatever external device was the source of the attack.

---

[40] See Halperin et al. (2008), p. 37.
[41] See Hei & Du (2011), p. 2.
[42] See Halperin et al. (2008).
[43] See Halperin et al. (2008).

VII. Securing a neuroprosthetic device vs. securing a neurocybernetic host-device system

A recurring theme throughout this text will be the distinction between ensuring information security for a neuroprosthetic device *per se* and ensuring information security for the larger neurocybernetic system that includes both the device and its human host. When discussing information security for neuroprosthetic devices, one must be careful to clarify whether the goal and effect of a particular security control is to strengthen the security of information contained within the device or within the larger host-device system.

For example, imagine a human being who possesses an advanced cochlear implant that records all of the person's auditory experiences and can later 'play back' any part of the recording internally through the person's cochlear nerve in a way that only he or she can hear, with the playback feature activated and controlled by acts of volition within the host's mind.[44] It may be the case that the cochlear implant possesses numerous security controls that make it almost impossible for an unauthorized party to directly access the information stored within it. Such controls might include, for example, an anti-tampering mechanism that destroys the device's internal memory if the device's physical casing is removed and a biometric control integrated into the device's processor that is based on the host's unique cognitive patterns and which disables the playback feature if the device were to be transplanted into some other host's body or physically connected to another computer. In this situation, one might say that all of the recorded auditory information stored on the device's internal memory is (almost) completely secure from access by any unauthorized party. But if we look not at the physical neuroprosthetic device itself but at the larger host-device system of which it is a component, we see that the information stored in the device is in fact highly unsecure: the device's host can play back recorded information (such as a conversation that he or she had overheard) in his or her mind through a simple act of will and then easily share that information with unauthorized parties simply by repeating it aloud, writing it down, or answering parties' questions about the information.

The device's host might share such information with unauthorized parties accidentally and unintentionally (e.g., sharing information about a sensitive conversation without realizing that the person with whom the information was being shared is an unauthorized party), as an intentional action performed by the host (e.g., sharing information from a damaging conversation in order to exact revenge on a disliked coworker), or under duress (e.g., as a

---

[44] Regarding the possibility of such playback devices, see Merkel et al., "Central Neural Prostheses" (2007), and Robinett, "The consequences of fully understanding the brain" (2002).

result of severe threats, blackmail, or enticement offered by the unauthorized party).

If the developers and operators of neuroprosthetic devices wish to maximally secure information contained within their devices, they must consider not only the characteristics and performance of a device as it exists in the abstract – physically and operationally separated from its human host – but also how the device functions when integrated into the neural circuitry of a particular human host. The host-device system may demonstrate unique cybernetic characteristics (such as feedback loops and other relationships of communication and control) that are neither visible nor even extant when the device and its host are considered separately or are, in fact, disconnected from one another.

Moreover, we will also extensively consider the possibility that an attack might be launched on a neuroprosthetic device by an adversary not for purposes of compromising information stored within the device itself but for the ultimate purpose of compromising the security of information stored within the natural biological systems and cognitive processes of the device's host (e.g., within the host's memory or conscious awareness) – perhaps by undermining his or her health or safety. There is also a possibility that an adversary could render information stored on a neuroprosthetic device damaged, disclosed, or inaccessible not by directly attacking the device but by launching an attack (whether by biochemical, physical, psychological, or other means) against the device's human host.

VIII. The weakest link – now at the heart of an information system

Human beings are considered to be the weakest link in any system of information security controls:[45] not only can we make unintentional physical or mental errors in operating a system or be fooled by social engineering attacks, we can also potentially become corrupted through greed, jealousy, resentment, lust, shame, pride, or ambition, and agree to take on an active and intentional role in disabling or bypassing our organization's security controls.[46]

With some kinds of information systems – e.g., those that are housed in physically and electronically isolated environments; are managed by a small and carefully screened team of expert personnel; contain no information that is financially, politically, or personally sensitive; and, once activated, perform

---

[45] See Sasse et al., "Transforming the 'weakest link'—a human/computer interaction approach to usable and effective security" (2001); Thonnard et al., "Industrial Espionage and Targeted Attacks: Understanding the Characteristics of an Escalating Threat" (2012); and Rao & Nayak, *The InfoSec Handbook* (2014), pp. 307-23.

[46] For the possibility of insider threats, see Coles-Kemp & Theoharidou, "Insider Threat and Information Security Management" (2010).

their tasks in an automated manner largely devoid of direct human intervention or control – the opportunity for security vulnerabilities to be intentionally exploited by human adversaries or unintentionally trigged by human agents can be reduced to a relatively small level. With other kinds of systems that have a much higher exposure to human activity – such as systems that are physically housed in publically accessible or mobile locations; have hundreds or thousands of individuals who possess privileged access to the system; and contain highly sensitive and valuable information that is provided, altered, and deleted daily by millions of human users utilizing web-based interfaces – the danger that a system's information security will eventually be compromised by human agents acting intentionally or unintentionally can be much greater.

It is easy to see that in the case of advanced neuroprosthetic devices, a system will always possess a large and crucial human element that cannot be eliminated: namely, the fact that the device is integrated into the neural circuitry of a human being. In this sense, a neuroprosthetic device inherently demonstrates a unique set of vulnerabilities that are found in no other systems, whether they be supercomputers, desktop computers, laptops, mobile or wearable devices, ubiquitous computing devices in smart homes or offices, web servers, automobiles, robotic manufacturing systems, communications satellites, video game systems, or any other computerized systems or devices.

The human mind – with its emotions, cognitive biases, incomplete knowledge, uneasy mix of gullibility and suspicion, and unique values and motivations – forms a perilous and unpredictable element of any information system. And in the case of an advanced neuroprosthetic device, that mind is often permanently anchored at the very heart of the system, where it is relentlessly active 24 hours a day in influencing and perhaps even controlling the functioning of the system – where it may be able to bring about some dramatic change in the contents of an information system or some tangible physical action in the world simply by means of an idle thought or volition or the recalling of a hazily outlined memory.

Many kinds of advanced neuroprostheses take the most dangerous and weakest possible link and embed it irrevocably at the very core of an information system which one nevertheless hopes to – somehow – make secure. Admittedly, this intimate connection between mind and machine can also possess its own unique advantages. The human mind tied to a neuroprosthetic device can display human strengths such as flexibility, intuition, the ability to correlate vast and unrelated pieces of knowledge and experience, creativity, and even faith, hope, and love – characteristics that allow a human mind not only to detect threats that a machine may be unable to recognize but also to wisely discern those rare instances when bypassing, disabling, or ignoring a security control in some particular circumstance may actually be the best (or only) way to ensure the information security of a system, individual, or organization. From the perspective of information security, integrating the neural circuitry of a human being and an electronic device can – for better or worse (or both) – bring with it not only the neurons and synapses of a human brain but also the intellectual, emotional, and even spiritual aspects of a human mind.

IX. New kinds of information systems to be protected – or used in executing attacks

Information security experts whose goal is to develop and implement mechanisms and procedures to ensure the information security of advanced neuroprosthetic devices and their corresponding host-device systems have a clear need to learn as much as possible about the capacities and uses of advanced neuroprosthetic devices. What may be less immediately obvious is that *all* InfoSec practitioners may need to develop at least some basic knowledge or awareness of the capacities and uses of advanced neuroprosthetic devices, insofar as such technologies provide powerful new tools by means of which attacks against all kinds of information systems – whether laptop computers, web servers, smartphones, archives of printed documents, or even human minds – can be launched and executed by sufficiently skilled adversaries.

Some adversaries may operate neuroprosthetic devices that are implanted in their own bodies. For example, a person could use an artificial eye to record secret video, an advanced cochlear implant to record conversations that are scarcely audible to a normal human ear, or an advanced virtual reality neuroprosthesis that allows him or her operate within cyberspace, sensing and manipulating it in a way that no unmodified human being could. Other adversaries might not host any neuroprosthetic devices within their own bodies, but they might be able to gain unauthorized access to neuroprosthetic devices implanted in other human hosts. For example, an adversary who hacked into the artificial eye of a corporation's vice president might use the ongoing live video feed to gain access to a plethora of financially valuable business secrets that are displayed on the vice president's computer screen and are contained in a corporate computer system that is otherwise impossible to break into. An adversary could gain access to a secured facility not by tunneling into the building but by hacking into the human security guard's mnemoprosthetic implant and creating a false memory of the 'fact' that the adversary is a senior staff member at the facility and should be welcomed when he arrives at the front door.

In such ways, the use of advanced neuroprostheses within human societies will require not only the development of a specialized subfield of information security dedicated to securing such devices and their host-device systems but also new approaches and responses across all of the other subfields of information security, as they adapt to the existence of such new technological means for planning and executing attacks on information systems.

X. New possibilities for biometrics

Other researchers have suggested that one approach to creating implantable devices that are highly secure during normal circumstances but that grant open access during health emergencies is to take advantage of one of the unique strengths of implantable devices: namely, their ability to draw rich, real-time biometric data from their human host. Biometrics that have been used for purposes of authentication for information systems in general (but not necessarily implantable neuroprostheses in particular) include:[47]

- Facial geometry
- Ear geometry
- Hand geometry (including vascular patterns)
- Fingerprints
- Palmprints
- Retinal patterns
- Iris patterns
- Infrared thermograms of patterns of heat radiated from the face or hand
- Signature and handwriting patterns
- Keystroke and typing patterns
- Gait and walking patterns
- Vocal characteristics
- Odor
- DNA

Many of these biometrics may be impractical if the designer of a neuroprosthetic device is attempting to create a security control mechanism within the device to ensure that it is still located within and being operated by its intended human host. For example, a memory implant located within its host's brain has no direct means by which to observe the iris patterns or patterns of heat radiated from the hands of its human host. However, an implanted neuroprosthetic device could potentially utilize some such biometrics if it had

---

[47] See Delac & Grgic, "A Survey of Biometric Recognition Methods" (2004), and Rao & Nayak (2014), pp. 297-303.

access to relevant biological sensory systems within its host: for example, if a sufficiently sophisticated neuroprosthetic device implanted within its host's brain had access to the optic nerve or visual cortex, it could conceivably conduct an iris scan by asking its host to stand in front of a mirror and look at the reflection of his or her own eyes. Instead of verifying that a neuroprosthesis is still implanted in its intended host, some such biometrics could potentially be used by a neuroprosthetic device to authenticate another individual (such as a maintenance technician) who is not the device's host but who is an authorized user and who should be given access to the device's systems: for example, an individual who possesses a bidirectional robotic prosthetic arm could potentially authenticate that another person is an authorized user simply by shaking the person's hand and thus detecting the person's hand geometry through touch or through optical or thermal sensors embedded in the prosthetic hand's palm or fingers.

Beyond such general-purpose biometrics, a number of biometrics have been developed or considered especially for use with implantable medical devices and are designed to take advantage of an implanted device's ability to directly access information about its host's internal biological processes. Such biometrics include:[48]

- Heart rate
- Breathing rate
- Blood glucose level
- Hemoglobin level
- Body temperature
- Blood pressure

Below we consider a number of biometrics and biometric systems that have been specifically proposed for or could conceivably be applied for use with advanced neuroprostheses.

A. Fingerprint type, eye color, height, and iris pattern

Hei and Du have proposed a biometric-based two-level security control to allow medical personnel to access implantable medical devices during an emergency. Prior to its implantation, a key is installed on an IMD that contains information about the basic fingerprint type, eye color, and height of its human host along with a code representing the host's iris pattern. When emergency medical personnel attempt to remotely access the IMD using their computer, the device will first ask the personnel to enter the host's fingerprint type, eye color, and height, as an initial access control; the medical personnel

---

[48] See Cho & Lee (2012), pp. 207-09.

can obtain all of this information by physically observing and manipulating the host, even if he or she is unconscious. As a more sophisticated control, the IMD then asks the medical personnel to take a photo of the host's iris (e.g., with a smartphone); an algorithm uses the photo to generate a code representing the iris pattern, which is then compared against the host's reference iris code stored on the IMD.[49]

Hei and Du note that such an approach would fail to provide access to legitimate medical personnel, for example, in cases in which the host's fingerprints had been damaged due to a fire or other injury or in which the host's body was trapped or positioned in such a way that the personnel could not photograph the host's eyes. Another possible disadvantage of this approach is the fact that it is based on the presumption that authorized emergency medical practitioners are the only individuals who would have both the desire to access and control the host's IMD and the ability to gather the necessary biometric data through direct physical interaction with the host. However, it seems possible that a sufficiently motivated adversary who wishes to gain unauthorized access to the host's IMD could potentially gather all of the needed biometric data simply by downloading high-quality images of the individual from the Internet.[50]

B. The heart's interbeat interval

Cho and Lee propose a model for secure communication among implanted biosensors or between an implanted biosensor and external system that uses the interbeat interval of the host's heartbeat.[51] The advantages of using that data source as a symmetric key for secure communications include the heartbeat's relatively high level of randomness (in comparison to some other potential biometrics) and the fact that the heart's interbeat interval can be detected by devices located throughout the host's body using a variety of mechanisms (e.g., by registering electrical activity or blood pressure) and can also be detected on the external surface of the host's body but cannot easily be detected by any adversary who does not have direct physical access to the host at that moment.

C. Strings generated from real-time ECG signals

Zheng et al. propose an "ECG-based Secret Data Sharing (ESDS) scheme" to secure information that is being transmitted from an implanted device to an external system. Before its transmission from the implanted device, data

---

[49] See Hei & Du (2011).

[50] For an example of this sort of vulnerability and risk, see Hern, "Hacker fakes German minister's fingerprints using photos of her hands" (2014).

[51] See Cho & Lee (2012).

is encrypted by the device using a key based on current biological activity that generates ECG signals that the device registers and which an external ECG is also capable of recording. After the message has been transmitted, it can be decrypted by an external system that had been using its own external ECG unit to record the host's activity at the same time and which can thus reconstruct the key.[52]

D. Gait and voice patterns

Vildjiounaite et al. propose a noninvasive multimodal model for securing personal mobile devices such as smartphones[53] that in principle could also be applied to implantable neuroprostheses. Their approach involves utilizing a device's built-in microphone and accelerometer to gather information about both the unique gait or walking patterns of the device's host along with unique characteristics of the host's voice. After activation, the security program enters a 'learning mode' for a period of a few days, during which time it records and analyzes the host's typical gait and voice patterns; if the device is able to establish suitably stable reference patterns, it then enters the 'biometric authentication mode' in which it regularly compares the gait and voice patterns that it is currently detecting against the reference patterns stored within it and – assuming that the current patterns and reference patterns match –authenticates the device's host and provides ongoing access to services.[54]

E. Behavior changes

Denning et al. note that one approach to securing implantable medical devices involves 'patient behavior changes,' in which the host of a device is asked to modify his or her behavior in some way as part of implementing a security control.[55] The sense in which Denning et al. use the phrase is broad enough to include cases in which the host's behavior change is the result or side-effect of a security control rather than the primary means by which the control is enforced. However, momentary 'behavior changes' could also be used as a sort of security control to verify that a neuroprosthetic device was still being used by its intended human host. For example, a device's host might periodically receive a certain kind of signal from the device, such as a visual alert displayed in the host's field of vision, an auditory alert produced

---

[52] See Zheng et al., "Securing wireless medical implants using an ECG-based secret data sharing scheme" (2014), and Zheng et al., "An ECG-based secret data sharing scheme supporting emergency treatment of Implantable Medical Devices" (2014).

[53] See Vildjiounaite et al., "Unobtrusive Multimodal Biometrics for Ensuring Privacy and Information Security with Personal Devices" (2006).

[54] Vildjiounaite et al. (2006), p. 197.

[55] Denning et al. (2010) p. 919.

through stimulation of the auditory cortex that the host can hear but no external parties can detect, or a stimulation of the host's proprioceptive system. After receiving the signal, the host then has a limited period of time in which to perform some particular (ideally inconspicuous) behavior that the implanted device can detect – such as blinking his or her eyes in a certain pattern, making a certain sound, or moving his or her fingers in a particular way. If the device detects the required behavior, it authenticates the device's host as an authorized party and allows ongoing access to the device's services for a particular period of time.

F. Thoughts and memories

One model of using a user's thoughts and memories as a biometric is presented by Thorpe et al. in their proposed mechanism for utilizing 'pass-thoughts.' In the simplest such approach, an authorized user memorizes a brief password, thereby storing it as a memory within his or her mind. When the user wishes to access a system, the system displays a random sequence of highlighted letters, and whenever the highlighted character happens to be the next character in the user's password, the user's brain generates a P300 potential spike (or positive potential that occurs roughly 300 milliseconds after the notable event) that can be detected by the system using an EEG or other device. Thorpe et al. note that such a pass-thought need not involve a text string; it could alternatively involve the use of "pictures, music, video clips, or the touch of raised pin patterns" or anything else that a person is capable of remembering and which the system is capable of displaying or presenting.[56] However, at its heart such a 'pass-thought' mechanism is essentially based on the use of a password as traditionally understood; the main difference is that rather than typing the password on a keyboard or speaking it aloud, an authorized user 'enters' the components of the password through interaction with a brain-computer interface that utilizes a device such as an EEG. Such a system thus displays many similarities with traditional password-based systems, insofar as an authorized user might forget his or her pass-thought (e.g., if it had been a long time since the user had last attempted to access the system); similarly, a user could be issued a new temporary pass-thought by a system's administrator (e.g., by displaying the contents of the pass-thought on a screen and asking the user to remember it) and the user could change the pass-thought to something new of his or her own choosing.

One can imagine other more sophisticated kinds of security controls based on cognitive processes such as thought and memory that are more exotic and which from an operational perspective have less in common with traditional password-based systems. For example, the same word (e.g., 'home'

---

[56] See Thorpe et al., "Pass-thoughts: authenticating with our minds" (2005).

or 'mother' or 'cat') or image (e.g., that of a wooded lake or a birthday cake) displayed to different individuals will generate different associations and the recall of different memories within each individual's mind because of the unique contents of each person's memories and life experience. The contents and internal interrelationships of such mental semantic networks could potentially be used as a form of authentication that cannot easily be lost, stolen, or spoofed, insofar as they are not directly accessible to parties outside of the mind that possesses them, and a human being cannot adequately understand and describe the nature and contents of his or her mental semantic networks even if he or she were intentionally attempting to do so (e.g., because the person were being subjected to threats or blackmail).

G. DNA

The use of DNA for verifying the identity of individuals has traditionally been limited to forensic applications rather than biometric access control for information systems, due to the fact that technologies have not yet been developed allowing simple real-time analysis and matching of a DNA sample with a reference pattern; however if such technologies were to someday be developed, DNA could potentially prove to be the most reliable of all biometrics (with some rare limitations, such as the case of identical twins whose DNA is indistinguishable).[57] Because of the uniqueness of DNA and its potentially high reliability as a biometric, some information security experts have suggested that it could someday be utilized as a biometric means for an implanted neuroprosthetic device to verify that it is operating within its intended human host.[58]

Spurred by ongoing advances in bionanotechnology, biomolecular computing, and related fields, DNA could also someday be used as a biometric or authenticator in other ways. For example, in the case of a neuroprosthesis that is composed of synthetic biological materials, other implanted systems could analyze the device's DNA in order to verify its origins and authenticate it (e.g., by locating a particular sequence of DNA within the device's genetic material in order to confirm that the device was created by the intended authorized manufacturer and was not an unauthorized growth or 'biohack' that had somehow been cultivated within the device's host through an adversary's introduction of engineered viruses, surgical nanorobots, or other unauthorized agents).

Multiple devices that have been implanted in the same host and which form a system could also potentially communicate with one another through

---

[57] Delac & Grgic (2004), p. 188.
[58] For example, the question of whether this might be a feasible approach has been posed by Pająk (2015).

their production of different viruses or biological material that are released into the host's bloodstream and travel between devices, e.g., utilizing the engineered virus's DNA or RNA as a data-storage mechanism for transmitting messages between the implanted devices and allowing one device to verify that the other devices are still present and functioning within the host's body. An electronic or biological neuroprosthetic device could also use DNA, for example, as a means of deploying and storing encryption keys for use in authenticating or being authenticated by other systems.[59]

H. Organismic continuity (or a continual 'liveness scan')

Qureshi notes that some kinds of biometric traits can be spoofed by presenting the biometric reader with an artificial construct of some sort that is not actually part of the authorized party's living organism but which nonetheless manifests patterns that mimic those of the person's organism.[60] Thus some kinds of fingerprint readers could potentially grant access to an adversary if the adversary presented a silicon 'finger' whose surface texture replicated the fingerprint pattern of an authorized user, and some kinds of iris scanners could potentially be fooled if presented a high-quality photograph of an authorized user's iris. Qureshi notes that one way to prevent such spoofing is to incorporate mechanisms for 'liveness detection' which verify that the presented biometric is actually being generated by a living organism.

For example, adding pulse and moisture detection capabilities to a fingerprint scanner can help the scanner to ensure that a presented biometric is being provided by a living finger and not a rubber replica, and an iris scanner could instruct its user to blink at certain moments, which induce predictable changes in the size of a living iris but not in a photographic replica.[61]

While helpful, such liveness detection is not foolproof. Even if an adversary were to possess sufficiently sophisticated genetic engineering and bioengineering technologies, it would generally not be possible for an adversary to directly 'grow' a living organ or body part capable of fooling a biometric scanner: for example, it is not possible to generate a replica of a human being's fingerprint or iris pattern simply by obtaining a sample of the person's DNA and attempting to culture a cloned finger or eyeball, since an individual's fingerprints and iris patterns are shaped by many environmental factors beyond simple genetics. However, if the details of an authorized user's fingerprints or iris pattern were known, a living replica could perhaps be created through

---

[59] For a discussion of the possibilities of using DNA as a mechanism for the storage of data, see Church et al., "Next-generation digital information storage in DNA" (2012).

[60] See Qureshi, "Liveness detection of biometric traits" (2011).

[61] Qureshi (2011), pp. 294-95.

other means – such as using a combination of nanotechnology and biotechnology to sculpt, reshape, or otherwise reengineer the finger or iris of an unauthorized living being so that its visible patterns sufficiently matched those of the authorized target.

However, in principle the concept of liveness detection could be applied to prevent such attacks and ensure that an implanted neuroprosthetic device is not only being accessed by *a* living being who displays certain characteristics but that it is being accessed by *the same living being* in whom it was originally implanted. For example, imagine that immediately upon its implantation in a human host, a neuroprosthetic device begins a continual, ongoing 'liveness scan' designed to ensure that the device is implanted in a living host – for example, by monitoring brain activity. Assuming that the process of scanning is reliable and uninterrupted, then as long as the scan has shown that from the moment of the device's implantation up to the present moment the monitored biological activity has continued without ceasing, then the device's software could be confident that the device is not only implanted within some living organism but within the organism of its original human host.[62]

Such a biometric security control based on the detection of 'organismic continuity' could be used, for example, to automatically disable – or delete the stored contents of – an implanted neuroprosthetic device upon the death of its human host or the cessation of particular brain activity within the host. Care would need to be given to the design of such systems to ensure that a neuroprosthetic device did not erroneously deactivate itself and cease to operate in cases in which its host had, for example, suffered a heart attack or stroke or entered a coma when such a termination of functionality was not the intention of the device's designer, manufacturer, or operator.

XI. Nontraditional computing platforms: from biomolecular computing and neural networks to nanorobotic swarms

When compared to conventional information systems such as desktop computers, laptops, and smartphones, an advanced neuroprosthetic device may be more likely to possess nonstandard, non-electronic components and to utilize nontraditional computing processes and formats. In the case of conventional computers, there is a decades-long history of design ingenuity, trial, error, and consumer feedback that has generated a body of experience and

---

[62] Such a model assumes that extracting the implant and transferring it to another living host could not be accomplished without at least a momentary break in the relevant biological activity recorded by the device; it also assumes that the device's design and structure are such that the recorded biological activity must actually be generated by biological activity occurring in biological material directly adjacent to the device and not, for example, spoofed through a targeted wireless transmission of certain types of electromagnetic radiation.

best practices allowing the efficient development and manufacturing of very powerful devices that utilize technologies such as silicon-based microprocessors, nonvolatile memory based in magnetic discs or flash memory, and computer programs constituting sets of instructions that can be loaded and executed by a central processing unit.

When developing new mass consumer electronics devices, it often makes more business sense for manufacturers to keep the cost and complexity of manufacturing processes at a minimum by designing devices that utilize well-established computing technologies while simultaneously attempting to advance those computing technologies in a way that incrementally enhances existing performance and capacities. When designing a next-generation mass-market smartphone intended for consumer release next year, it would most likely be seen as an unnecessarily exotic (and practically and economically unfeasible) approach for a manufacturer to attempt to build the device's internal computer on a platform utilizing biomolecular computing, quantum computing, or physical neural networks.

Advanced neuroprosthetic devices, on the other hand, already inherently incorporate and rely on at least some highly 'exotic' and 'nonstandard' computing components and processes, insofar as they must integrate both physically and operationally with the biological structures and neural circuitry of their human host. When considering information security for advanced neuroprosthetic devices, one cannot assume that a neuroprosthetic device will be a traditional computing device – with traditional kinds of components, architectures, memory systems, and ways of gathering and processing information to generate actions and output – that has simply been implanted into the body of a human host.

While some advanced neuroprostheses might indeed resemble a smartphone that has just been miniaturized and implanted in a human body, other neuroprosthetic devices might scarcely be recognizable as computers – or even technological devices. Neuroprostheses that perform the processing of information by means of a physical neural network might be partially or fully constructed from biological materials and may be integrated into the body of their host in a way that makes it difficult to discern – both structurally and operationally – where the host ends and the device begins. Information security might involve protecting a neuroprosthetic device not only against computer viruses but against biological viruses, as well. In order to avoid invasive surgery that could damage a human host's brain, other neuroprosthetic devices might consist of a swarm of nanorobots that have been designed to be capable of crossing the blood-brain barrier and which are introduced into the host's bloodstream and find their way to the correct location in the brain, where they work together to stimulate (or are stimulated by) the brain's natural interneurons in particular ways, even while retaining their physically diffuse structure.[63]

The possibility that neuroprosthetic devices might take such forms creates unique issues and considerations for information security. On the one hand, the use of nontraditional components, structures, and computing methods may render a neuroprosthetic device more secure, because common kinds of attacks that are often effective against conventional computers and information systems may be ineffective or even wholly inapplicable in the case of neuroprosthetic devices. On the other hand, the use of nontraditional elements may mean that the designers, manufacturers, operators, and hosts of neuroprosthetic devices cannot rely on the vast body of information security knowledge and best practices that have been developed over decades for securing conventional computer systems, because many of those information security strategies, mechanisms, and techniques may also be ineffective or inapplicable in the case of neuroprostheses.

XII. Technology generating posthuman societies and posthuman concerns

'Posthumanism' can be defined as a conceptual framework for understanding reality that is post-anthropocentric and post-dualistic; it views the 'natural' biological human being as traditionally understood as just one of many intelligent subjects acting within the world's complex social ecosystem.[64] Some forms of posthumanism explore the historical ways in which our notion of typical human beings as the only members of society has been perpetually challenged by the generation of cultural products like myths and literary works that feature quasi-human beings such as monsters, ghosts, angels, anthropomorphic animals, cyborgs, and space aliens (or in other words, through processes of *nontechnological posthumanization*).[65] Other forms of

---

[63] See Al-Hudhud, "On Swarming Medical Nanorobots" (2012).

[64] This definition builds on the definitions formulated by scholars of posthumanism such as Ferrando, Miller, Herbrechter, Miah, and Birnbacher. See Ferrando, "Posthumanism, Transhumanism, Antihumanism, Metahumanism, and New Materialisms: Differences and Relations" (2013), p. 29; Miller, "Conclusion: Beyond the Human: Ontogenesis, Technology, and the Posthuman in Kubrick and Clarke's 2001" (2012), p. 164; Herbrechter, *Posthumanism: A Critical Analysis* (2013), pp. 2-3; Miah, "A Critical History of Posthumanism" (2008), p. 83; and Birnbacher, "Posthumanity, Transhumanism and Human Nature" (2008), p. 104, as well as the typology of posthumanism formulated in Part One of Gladden, *Sapient Circuits and Digitalized Flesh: The Organization as Locus of Technological Posthumanization* (2016).

[65] Such forms of posthumanism include the critical and cultural posthumanism pioneered by Haraway, Halberstam and Livingstone, Hayles, Badmington, and others. See, e.g., Haraway, "A Manifesto for Cyborgs: Science, Technology, and Socialist Feminism in the 1980s" (1985); Haraway, *Simians, Cyborgs, and Women: The Reinvention of Nature* (1991); *Posthuman Bodies*, edited

posthumanism investigate the ways in which the circle of persons and intelligent agents that constitute our social ecosystem is being transformed and expanded through the engineering of new kinds of entities such as human beings possessing neuroprosthetic implants, genetically modified human beings, social robots, sentient networks, and other advanced forms of artificial intelligence (i.e., through processes of *technological posthumanization*).[66] The development of rigorous and insightful forms of posthumanist thought is becoming increasingly important, as society grapples with the ontological, ethical, legal, and cultural implications of emerging technologies that are generating new forms of posthumanized existence.

Philosophers of technology have given much thought to the transformative effects that technology can play in the lives of human beings, either as a means of liberation and self-fulfillment or as a source of oppression and dehumanization.[67] Within the context of posthumanization, neuroprosthetic devices are expected to increasingly become gateways that allow their human hosts to more deeply experience, control, and be controlled by the structures and dynamics of such digital-physical ecosystems.[68] Advanced neuroprostheses thus have the potential to reshape human psychological, social, and cultural realities – and even challenge many popular notions of what it means to be 'human' – in ways greater and more powerful than those demonstrated by previous generations of technology. Already 'cyborg-cyborg interaction' is becoming a fundamental aspect of society,[69] and genetic engineering may accelerate trends of 'cyborgization' by further enhancing the ability of the human

---

by Halberstam & Livingstone (1995); Hayles, *How We Became Posthuman: Virtual Bodies in Cybernetics, Literature, and Informatics* (1999); Graham, *Representations of the Post/Human: Monsters, Aliens and Others in Popular Culture* (2002); Badmington, "Cultural Studies and the Posthumanities" (2006); and Herbrechter (2013).

[66] Such forms of posthumanism include philosophical posthumanism, bioconservatism, and transhumanism, which are analyzed in Miah (2008), pp. 73-74, 79-82, and Ferrando (2013), p. 29. Such approaches can be seen, for example, in Fukuyama, *Our Posthuman Future: Consequences of the Biotechnology Revolution* (2002); Bostrom, "Why I Want to Be a Posthuman When I Grow Up" (2008); and other texts in *Medical Enhancement and Posthumanity*, edited by Gordijn & Chadwick (2008).

[67] For a discussion of such questions in the context of human augmentation and neuroprosthetics, see Abrams, "Pragmatism, Artificial Intelligence, and Posthuman Bioethics: Shusterman, Rorty, Foucault" (2004); Kraemer, "Me, Myself and My Brain Implant: Deep Brain Stimulation Raises Questions of Personal Authenticity and Alienation" (2011); Erler, "Does Memory Modification Threaten Our Authenticity?" (2011); Tamburrini, "Brain to Computer Communication: Ethical Perspectives on Interaction Models" (2009); and Schermer, "The Mind and the Machine. On the Conceptual and Moral Implications of Brain-Machine Interaction" (2009).

[68] See Gladden, "Neural Implants as Gateways to Digital-Physical Ecosystems and Posthuman Socioeconomic Interaction" (2016).

[69] See Fleischmann, "Sociotechnical Interaction and Cyborg–Cyborg Interaction: Transforming the Scale and Convergence of HCI" (2009).

body to interface at a structural level with implanted or external technological systems. Posthumanizing neuroprosthetics will allow for increasingly intimate forms of communication that do not involve physical face-to-face interaction but are instead mediated by technology, thereby facilitating the development of new kinds of posthuman interpersonal relationships and social structures.[70] In a sense, then, questions about the information security of advanced neuroprosthetic devices and host-device systems should be considered in a broader context relating to human societies and the human species: future posthumanizing neuroprostheses should be developed (or, if appropriate, *not* developed) in such a way that will ensure not only that individual persons' information security can be protected – but that a humanity and human species can continue to exist whose members are able to generate, use, and exchange information with confidence in its security.

XIII. Human autonomy, authenticity, and consciousness: risk management and the possibility of ultimate loss

Posthumanizing neuroprostheses thus force us to ponder the future existence and nature of humanity in a way that is elicited by other technologies such as nuclear weaponry or genetic engineering but is not found, for example, with technologies such as desktop computers or even self-driving automobiles. Nevertheless, from the perspective of information security, the primary focus with regard to advanced neuroprosthetic devices is typically very much the technology's impact on and use by particular human beings; and through their interaction with a human being's neural circuitry, many advanced neuroprostheses have the potential to reshape and transform an individual life in ways that are incredibly powerful – and can be either beneficial or harmful.

One of the gravest (and most unique) concerns that an information security professional must consider with regard to neuroprosthetic devices is the impact that they might have on the autonomy, authenticity, and conscious awareness of their human host.[71] Researchers have found, for example, that

---

[70] See Fleischmann (2009) and Grodzinsky et al., "Developing Artificial Agents Worthy of Trust: 'Would You Buy a Used Car from This Artificial Agent?'" (2011).

[71] For an exploration of the ways in which the implantation and use of advanced neuroprosthetic devices (and the accompanying process of 'cyborgization' of the devices' human hosts) can contribute to a new form of personal identity for a host-device system that fuses both biological, cultural, and technological elements, see Kłoda-Staniecko, "Ja, Cyborg. Trzy porządki, jeden byt. Podmiot jako fuzja biologii, kultury i technologii" (2015). For a discussion of the significance of the physical boundaries of a human organism and the ways in which technologies such as implantable neuroprostheses can impact cognitive processes and the ways in which a person is

some human beings who have utilized neuroprosthetic devices for deep brain stimulation in order to treat conditions such as Parkinson's disease have reported feelings of reduced autonomy and authenticity: some such individuals find it impossible to know any longer whether 'they' are actually the ones responsible for their thoughts, desires, emotions, and decisions, or whether these mental phenomena are being influenced, controlled, or even created by the electrodes firing deep within their brains.[72]

It is possible to imagine a concrete outcome of the use of particular kinds of (poorly designed or implemented) neuroprosthetic devices which – from the perspective of their human hosts – would produce not only harm but the termination of their personal identity and annihilation of their existence as a human subject within the world. For example, a mnemoprosthetic implant that is designed to enhance its users' memory capacities but which causes some of its users to enter a coma or vegetative state would be legally and ethically impermissible – not to mention being counterproductive from an information security perspective, insofar as it would render the information contained within such a user's mind unavailable even to that user himself or herself.

Arguably, though, an even worse scenario would be that of a neuroprosthetic device that permanently destroys the autonomy, consciousness, personal identity, continuity of sapient self-awareness, and metavolitionality (or conscience) of its human host – in other words, that obliterates the 'essence' of what makes that person human – but that does so in such a way that this destruction is not detectable to other human beings. For example, consider an extremely sophisticated neuroprosthetic device consisting of a vast network of nanorobotic components that occupy interstitial spaces within the brain and are designed to support the synaptic activity of individual neurons.[73] Imagine that – in a manner that may not be recognized or understood

---

understood as a moral subject or subject of experiences, see Buller, "Neurotechnology, Invasiveness and the Extended Mind" (2011). For a philosophical analysis of the ways in which personal autonomy is threatened by brain-machine interfaces, see Lucivero & Tamburrini, "Ethical Monitoring of Brain-Machine Interfaces" (2007). Questions of personal identity and authenticity are explored by Schermer (2009).

[72] See Kraemer (2011) and Van den Berg, "Pieces of Me: On Identity and Information and Communications Technology Implants" (2012). It should be noted that Kraemer observes that other users of neuroprostheses for deep bran stimulation have reported precisely the opposite experience: they feel as though the neuroprosthetic devices have *restored* their autonomy and given them increased authenticity as – for the first time in years – they are in control of their bodies once again.

[73] Such technologies have been proposed by some transhumanists as a possible path toward 'mind uploading.' See Koene, "Embracing Competitive Balance: The Case for Substrate-Independent Minds and Whole Brain Emulation" (2012); Proudfoot, "Software Immortals: Science or Faith?" (2012); Pearce, "The Biointelligence Explosion" (2012); Hanson, "If uploads come first: The

even by the device's designers – this neuroprosthetic device does not actually 'support' the natural synaptic activity of the brain's biological neurons but instead controls or replaces it. The physical synaptic connections between neurons (and thus their communication) are disrupted and replaced by connections between the nanorobotic 'pseudo-neurons.' The person's physical body can still react to environmental stimuli, walk, smile, and even engage in meaningful conversations with family or friends – but in fact, the person's conscious awareness has been eliminated and the person has undergone a sort of 'brain death'; all of the outward physical activity is simply being orchestrated by the extremely sophisticated processes of the artificial neural network or computer program that controls the nanorobotic system and which causes it to stimulate particular motor neurons at a particular time in order to generated desired motor behaviors. In effect, the person's body has become a sort of neurocybernetic 'zombie,' a mindless puppet controlled by the puppeteer of the neuroprosthetic device.[74]

There are some transhumanists (e.g., proponents of the idea of 'mind uploading') who might argue that such a device would not truly destroy the consciousness or essence of its human host – and that even if it did, they would be willing and even eager to transform their own bodies through the use of such a device, insofar as it might provide a bridge that would allow them to 'transfer' their memories and patterns of mental activity into a robotic or computerized body that would essentially allow them, as they see it, to live forever. There may indeed be human beings who would be happy to imagine that at the cost of destroying their own embodied consciousness, a biomechanical automaton or robot could be created that would go about its activities in the world, replicating the habits and behaviors and continuing the social relationships of the individual who had served as its template, simulating that person's emotions and recreating his or her memories in the way that a video recording recreates some filmed event. But presumably most human beings would consider a neuroprosthetic device that destroys their embodied conscious awareness to be an absolutely impermissible – and, indeed, lethal – outcome, regardless of whatever other effects it might yield.

A dilemma for information security professionals (and the designers and operators of neuroprosthetic devices) is that in principle it may sometimes be impossible to know what effect a neuroprosthesis is truly having on the cognitive processes – and especially, on the lived conscious experience – of its human host. If the human host of an experimental neuroprosthesis asserts

---

crack of a future dawn" (1994); and Moravec, *Mind Children: The Future of Robot and Human Intelligence* (1990), for a discussion of such issues from various perspectives.

[74] For a discussion of such possibilities, see Gladden, *Neuroprosthetic Supersystems Architecture* (2017), pp. 133-34.

that the implantation and activation of the device has in no way harmed or diminished his or her sapience and conscious awareness, this could mean that the device has indeed had no such effect – or that it has destroyed the host's conscious awareness and agency, and the source behind that statement was not the human being but the agency of the neuroprosthetic device itself. It is also possible to imagine a situation in which the conscious awareness and autonomous agency of the human host might still exist – but no longer has the ability to control the motor systems of its body and alert the outside world to the fact that it is essentially 'trapped' helplessly within a body whose sensorimotor systems are now controlled by the neuroprosthesis.[75] Although more sophisticated forms of neural scanning and imaging, computer simulations of the effects of neuroprosthetic devices, research into artificial intelligence, and philosophical thought experiments may be able to provide one with reasonable grounds for suspecting (or doubting) that such a situation is possible, it may be difficult to definitively exclude the possibility if there are no independent means for settling the question outside of internal conscious experience (or lack thereof) of a device's human host.

However remote they might be, such possibilities create particular challenges for risk management, insofar as one must grapple with the danger of an occurrence whose probability of being realized appears quite small (but which, in fact, cannot be reliably determined) and whose nightmarish effect on a device's host would, if realized, be lethal – if not worse. While philosophers and other researchers have begun to seriously debate such issues (especially with regard to the technological and ontological feasibility of mind uploading[76]), deeper exploration of such issues from metaphysical, psychological, ethical, legal, and even theological perspectives is required.[77]

---

[75] In a sense, such an occurrence would be the (unfortunate) mirror opposite of those positive situations in which a neuroprosthetic device provides the only means of communication with the outside world for locked-in patients who are completely paralyzed yet fully conscious, including those suffering from ALS, stroke, or traumatic brain injury. For a discussion of such positive cases, see Donchin & Arbel, "P300 Based Brain Computer Interfaces: A Progress Report" (2009).

[76] See, e.g., Koene (2012); Proudfoot (2012); Pearce (2012); Hanson (1994); and Moravec (1990).

[77] For a discussion of many ethical issues relating to neuroprosthetics, see Iles, *Neuroethics: Defining the Issues in Theory, Practice, and Policy* (2006). For an explicit consideration of ethical issues in light of information security concerns (and the possibility that adversaries could potentially wish to gain access to neuroprosthetic devices), see Denning et al. (2009). For theological and spiritual issues relating to neuroprosthetic devices, see Campbell et al., "The Machine in the Body: Ethical and Religious Issues in the Bodily Incorporation of Mechanical Devices" (2008).

XIV. The nexus of information security, medicine, biomedical engineering, neuroscience, and cybernetics

One aspect of information security that is highlighted by our considera-tion of advanced neuroprosthetic devices is the growing relationship of infor-mation security to fields such as medicine, biomedical engineering, and neu-roscience – and the importance that the knowledge developed in these fields will have for shaping the future of information security.[78]

It is already the case that information security is a transdisciplinary field in which personnel must not only be experts in computer hardware and soft-ware but must also have knowledge of fields such as psychology, finance, law, and ethics. However, the growing use of neuroprosthetic devices will mean that information security personnel will also need to possess at least basic knowledge about the biological and neuroscientific aspects of such devices. Some large organizations may even find it desirable and feasible to add to their information security teams physicians, neuroscientists, and biomedical engineers who can work with the other team members to ensure, for example, that any information security mechanisms or practices that the organization implements in relation to its employees' neuroprosthetic devices do not re-sult in biological or psychological harm to the employees. Such medical ex-pertise would also be necessary in order for information security personnel to design safe and effective countermeasures that can be employed against adversaries who possess their own neuroprostheses and attempt to employ them to carry out acts of illicit surveillance or corporate espionage against the company. By employing a knowledge of biology, biomedical engineering, and neuroscience, an organization's information security personnel could de-velop security controls and countermeasures that neutralize such threats without causing biological or psychological injury to suspected adversaries for which the company and its information security personnel could poten-tially be legally and ethically responsible and financially liable.[79]

One challenge that arises in attempting to link information security with medicine is that the two fields utilize different vocabularies and conceptual frameworks: information security is grounded largely in the theoretical framework of computer science while medicine is rooted in that of biology and chemistry. In addressing this challenge, it may be helpful to build on the

---

[78] For the increasingly inextricable connections between medical devices and information tech-nology, see Gärtner, "Communicating Medical Systems and Networks" (2011).

[79] For a related discussion of questions about the legality and ethicality of undertaking offensive countermeasures against botnets, see Leder et al., "Proactive Botnet Countermeasures: An Of-fensive Approach" (2009).

field of cybernetics, which was founded to provide precisely the sort of trans-disciplinary theoretical framework and vocabulary that can be used to translate insights between all of the fields that study patterns of communication and control – whether it be in machines, living organisms such as human beings, or social systems.[80]

Drawing on such diverse manifestations of cybernetics as biocybernetics, neurocybernetics, and management cybernetics, it may be possible to envision the human brain, its surrounding body, and any neuroprosthetic devices, implantable computers, and other internal or external technological systems that are integrated into the body as together forming a single physical 'shell' for the human mind connected with that body.[81] The human brain, body, and technological devices together constitute a system that receives information from the external environment, processes, stores, and utilizes information circulating within the system, and transmits information to the external environment, thereby creating networks of communication and control. In such a model, information security experts, physicians, and biomedical engineers would thus share the single task of ensuring the secure, productive, and effective functioning of this entire information system that may contain both biological and electronic components – with that common goal only being achievable if all of the expert personnel involved succeed in fulfilling their unique individual roles.

XV. Conclusion

Thanks to decades of tireless labor by researchers and practitioners, there now exists a coherent body of knowledge and best practices relating to information security for computerized information systems that is well-developed and battle-tested and which is being continually refined to deal with new kinds of threats. While those experts who will strive to provide information security for advanced neuroprostheses will be able to ground their efforts in the existing practice of InfoSec for computerized information systems, that general body of knowledge will, on its own, prove to be an inadequate source and guide for their efforts – because advanced neuroprosthetic devices are not simply computerized information systems. In many cases, an advanced neuroprosthetic device simultaneously possesses at least three different natures; it combines in a single device (1) a computerized information system with (2) an implantable medical device and (3) a posthumanizing technology that has the potential to transform the mind of its human host and radically reshape its user's relationship with his or her own mind and body, with other

---

[80] See Wiener, *Cybernetics: Or Control and Communication in the Animal and the Machine* (1961).

[81] Such a perspective might be understood as comprising a sort of 'celyphocybernetics,' from the Ancient Greek κέλυφος, meaning 'shell,' 'sheath,' or 'pod.'

human beings, with technological systems, and with the external environment as a whole.

Taking into account all of the issues that we have considered earlier in this chapter, it becomes apparent that practices and mechanisms designed to protect the information security of generic computerized information systems are insufficient – if not irrelevant or, in some cases, even counterproductive – when it comes to protecting the information security of advanced neuroprosthetic devices and their host-device systems.[82] As a result, many existing neuroprosthetic devices do not incorporate adequate security controls and do not sufficiently protect the privacy of their human hosts and users.[83]

We would argue that in order to implement robust and effective approaches for advancing the information security of advanced neuroprostheses as information systems, medical devices, and transformative posthumanizing technologies, new conceptual frameworks will first need to be explicitly developed. Such frameworks include **device ontologies** that help one to identify and describe the relevant characteristics of a neuroprosthetic device in a systematic manner; **typologies** that use the ontologies to categorize different neuroprosthetic devices into groups that possess similar relevant characteristics; and neuroprosthetic security **protocols** that define specific device characteristics and operational practices that should be implemented in particular circumstances, based on the needs of a device's host and operator and the broader context of the device's use (including legal, ethical, and organizational considerations). Several such conceptual frameworks are presented in the companion volume to this text, *Neuroprosthetic Supersystems Architecture: Considerations for the Design and Management of Neurocybernetically Augmented Organizations;* another is presented in the following chapter. While these are designed primarily to address the unique circumstances of advanced neuroprostheses, they may also yield insights that can be adapted for promoting the information security of a broader array of future 'neurotech' and its human users.

---

[82] Regarding, e.g., the need for new regulatory frameworks relating to implanted ICT devices, see Kosta & Bowman, "Implanting Implications: Data Protection Challenges Arising from the Use of Human ICT Implants" (2012). For an example of the complexities involved with determining which regulations and standards apply to which kinds of medical systems and devices, see Harrison, "IEC80001 and Future Ramifications for Health Systems Not Currently Classed as Medical Devices" (2010). For the inadequacy of traditional information security frameworks as applied to e-healthcare in general, see Shoniregun et al., "Introduction to E-Healthcare Information Security" (2010).

[83] See Tadeusiewicz et al., "Restoring Function: Application Exemplars of Medical ICT Implants" (2012).

Chapter Four

A Two-dimensional Framework of Cognitional Security for Advanced Neuroprosthetics

Abstract. In this text a two-dimensional 'cognitive security' framework is developed for advanced neuroprosthetic devices that takes into account not only the information security needs of a neuroprosthesis itself but also those of the host-device system that the device creates through its integration into the neural circuitry of its human host. The framework first describes nine InfoSec goals or attributes – namely, confidentiality, integrity, availability, possession, authenticity, utility, distinguishability, rejectability, and autonomy. The framework considers how the pursuit of these goals for a host-device system can be advanced (or subverted) at three different levels, in which the human host of a neuroprosthetic device is considered in his or her role as: 1) a sapient metavolitional agent; 2) an embodied embedded organism; and 3) a social and economic actor. This framework shares some common elements with classical models of InfoSec goals that were formulated for general-purpose computing and information systems, but it also proposes new elements to address the unique nature of advanced neuroprostheses.

I. Developing a cognitional security framework

In this chapter we develop a two-dimensional conceptual framework for cognitional security. The first dimension includes nine essential information security attributes or goals for neuroprosthetic devices and host-device systems, namely **confidentiality**, **integrity**, **availability**, **possession**, **authenticity**, **utility**, **distinguishability**, **rejectability**, and **autonomy**. Each of these attributes relates to the host-device system as understood at three different levels, which constitute the second dimension of the framework; the levels are those of a device's host understood as **sapient metavolitional agent**, **embodied embedded organism**, and **social and economic actor**. Below we present this framework in detail and consider its implications for information security for advanced neuroprosthetic devices.

II. Defining security goals for the entire host-device system: nine essential attributes

One of the most fundamental ways of conceptualizing information security is through a framework of essential characteristics that a system must possess in order to be secure. One can understand these characteristics as the security 'attributes' that an ideal system would possess. However, in practice such characteristics can never be perfectly achieved, and thus rather than envisioning them as a system's optimal state, they can instead be understood as the security 'goals' that one is perpetually striving to attain through the *process* of information security.

Denning et al. propose a model of 'neurosecurity' for neuroprosthetic devices that strives for "the protection of the confidentiality, integrity, and availability of neural devices from malicious parties with the goal of preserving the safety of a person's neural mechanisms, neural computation, and free will."[1] While that model provides an excellent starting point (especially with regard to contemporary types of neuroprosthetic devices that are already in use), in itself it is not sufficiently specific or robust to drive the development of mature and highly effective InfoSec plans, mechanisms, and practices that will be capable of protecting neuroprosthetic devices and host-device systems from the full range of threat sources, including expertly skilled and intensely motivated adversaries. In particular, a stronger and more comprehensive information security framework will be needed to protect the kinds of highly sophisticated (and even posthumanizing) neuroprosthetic devices and host-device systems that are expected to become a reality within the coming years and decades.

The cognitional security framework that we formulate here for a host-device system utilizing advanced neuroprosthetics includes nine security goals or attributes: three are the elements of the classic CIA Triad (confidentiality, integrity, and availability);[2] three are additional characteristics developed by Donn Parker in his security hexad (possession, authenticity, and utility);[3] and

---

[1] See Denning et al., "Neurosecurity: Security and Privacy for Neural Devices" (2009).

[2] Rao & Nayak, *The InfoSec Handbook* (2014), pp. 49-53.

[3] See Parker, "Toward a New Framework for Information Security" (2002), and Parker, "Our Excessively Simplistic Information Security Model and How to Fix It" (2010). There is ongoing debate within the field of information security regarding the number and relationship of InfoSec goals and attributes. Other attributes identified by some, such as 'completeness' and 'non-repudiation/accuracy' (see Dardick's analysis of IQ, CFA, and 5 Pillars in Dardick, "Cyber Forensics Assurance" (2010)) or 'accountability' and 'assurance' (see *NIST Special Publication 800-33: Underlying Technical Models for Information Technology Security* (2001), p. 3) are not explicitly considered here as independent objectives.

three are new characteristics which we have identified as being uniquely relevant for the security of advanced neuroprosthetics (distinguishability, rejectability, and autonomy).[4] Below we briefly define each of these security goals, with particular reference to their relevance for advanced neuroprosthetics.

A. Confidentiality

In the context of an advanced neuroprosthetic system, we can define confidentiality as "limiting the disclosure of information to only those sapient agents that are authorized to access it." Note that according to this understanding, confidentiality has only been breached if the information is accessed by another 'sapient agent' (such as a human being) who is not authorized to do so. For example, imagine that a neuroprosthesis implanted in your brain is able to detect and record the contents of your thoughts and – without your knowledge – is wirelessly transmitting a record of this data to create a 'backup copy' of your thoughts on an external computer. While this means that you no longer have sole control or 'possession' of the information (as defined below), the creation of such an unauthorized external backup of your thoughts does not in itself represent a loss of confidentiality, as long as the information stored on the external computer is not viewed by some person not authorized by you.

Our definition of confidentiality in relation to neuroprostheses builds on existing definitions used in the field of information security. For example, confidentiality has previously been defined as "the requirement that private or confidential information not be disclosed to unauthorized individuals. Confidentiality protection applies to data in storage, during processing, and while in transit."[5] Parker defines confidentiality as the "Limited observation and disclosure of knowledge."[6] Alternatively, it can be understood as "Preserving authorized restrictions on information access and disclosure, including means for protecting personal privacy and proprietary information."[7] Dardick proposes a model of Cyber Forensics Assurance (CFA) in which confidentiality is understood as "ensuring that information is accessible only to

---

[4] For a discussion of the characteristics of distinguishability, rejectability, and autonomy in the context of implantable cognitive neuroprostheses, see Gladden, "Information Security Concerns as a Catalyst for the Development of Implantable Cognitive Neuroprostheses" (2016).

[5] *NIST SP 800-33* (2001), p. 2.

[6] Parker (2002), p. 125.

[7] 44 U.S.C., Sec. 3542, cited in *NIST Special Publication 800-37, Revision 1: Guide for Applying the Risk Management Framework to Federal Information Systems: A Security Life Cycle Approach* (2010), p. B-2.

those authorized to have access."[8] Confidentiality applies not only to the data stored within a system but also to information about the system itself,[9] insofar as knowledge about a system's design, functioning, and vulnerabilities makes it easier for unauthorized parties to plan an attack on the system.

While ensuring that information is not disclosed to unauthorized parties is typically an important organizational goal, preventing the destruction or corruption of the information is often an even more important objective. Thus *NIST SP 800-33* notes that "For many organizations, confidentiality is frequently behind availability and integrity in terms of importance. Yet for some systems and for specific types of data in most systems (e.g., authenticators), confidentiality is extremely important."[10] As is true for implantable medical devices generally, neuroprosthetic devices constitute a class of systems whose data is often highly sensitive and for which confidentiality is thus a great concern.

B. Integrity

With regard to an advanced neuroprosthetic system, we can define integrity as "remaining intact, free from the introduction of substantial inaccuracies, and unchanged by unauthorized manipulation."

As is true for confidentiality, integrity is needed for both the data stored within a system as well as for the storage system itself.[11] The integrity of information in advanced neuroprosthetic systems is a complex issue, especially in the case of neuroprostheses that are involved with the mind's processes for forming, storing, and recalling memories. The long-term memories that are stored within our brain's natural memory systems already undergo natural processes of compression and degradation over time;[12] none of them contains a perfect representation of the original experience that led to formation of the memory. While our memories may, over time, lose detail and become more impressionistic, they do not lose 'integrity' unless a memory has been transformed in such a way that the meaning that it *does* convey is no longer accurate or no longer presents a coherent whole. According to our definition, a memory also does not lose integrity simply as a result of undergoing manipulation, as long as it is a form of *authorized* manipulation that does not

---

[8] Dardick (2010), p. 61. Dardick developed his CFA model by analyzing and synthesizing definitions developed in frameworks such as the CIA Triad as defined in the Federal Information Security Management Act of 2002 (FISMA), the Five Pillars of Information Assurance model developed by the US Department of Defense, the Parkerian Hexad, and the Dimensions of Information Quality developed by Fox and Miller.

[9] *NIST SP 800-33* (2001), p. 2.

[10] *NIST SP 800-33* (2001), p. 2.

[11] *NIST SP 800-33* (2001), p. 2.

[12] See Dudai, "The Neurobiology of Consolidations, Or, How Stable Is the Engram?" (2004).

introduce substantial inaccuracies. (Thus a neuroprosthetic device that uses some algorithm to compress memories by identifying and preserving essential details while eliminating inessential elements would not necessarily be damaging the integrity of those memories.)

Within more generalized existing frameworks for information security, data integrity has been defined as "the property that data has not been altered in an unauthorized manner while in storage, during processing, or while in transit," and system integrity has been defined as "the quality that a system has when performing the intended function in an unimpaired manner, free from unauthorized manipulation."[13] It is alternatively understood as "Guarding against improper information modification or destruction, and includes ensuring information non-repudiation and authenticity."[14] Parker defines integrity as the "Completeness, wholeness, and readability of information" and the fact that the information remains "unchanged from a previous state."[15] Dardick's synthetic CFA model summarizes the joint concept of "Integrity/Consistency" as the "perceived consistency of actions, values, methods, measures and principle – unchanged 'is it true all of the time?' (Verification)."[16] *NIST SP 800-33* suggests that after availability, integrity is frequently an organization's most important InfoSec goal.[17]

C. Availability

In the context of an advanced neuroprosthetic system, we can define availability as **"the ability to access and experience desired information in a timely and reliable manner."** This definition of availability differs somewhat from definitions traditionally used in information security. First, it emphasizes that for the user of a neuroprosthetic system, it is not sufficient for information to be stored in a database from which the user can export or save files with particular subsets of information; it is typically important that the user be able to directly *experience* the information as an object of his or her conscious awareness (e.g., sense data that are presented to one's mind to be perceived in the form of percepts or memories that can be recalled and thus 're-experienced' in one's mind at will). Second, this definition emphasizes that it is not sufficient for a user to have access to a vast pool of information in which the one or two pieces of information that the user would actually like to consciously recall are lost amidst countless streams of information, most of which are at the moment irrelevant. The user of a neuroprosthetic device must be able to

---

[13] *NIST SP 800-33* (2001), p. 2.
[14] 44 U.S.C., Sec. 3542, cited in *NIST SP 800-37* (2010), Rev. 1, p. B–6.
[15] Parker (2002), p. 125.
[16] Dardick (2010), p. 61.
[17] *NIST SP 800-33* (2001), p. 2.

quickly and reliably experience in his or her conscious awareness the partic-ular piece of information that he or she desires. In the case of some neuro-prostheses, such as an artificial eye that is conveying sense data from the en-vironment, 'quickly' experiencing information effectively means that it must be presented in real time.

Ensuring the availability of information involves maintaining both data and the system or systems that contain it and provide it to users. More gen-eralized frameworks for information security have defined availability as the assurance "that systems work promptly and service is not denied to author-ized users;" it involves preventing any "unauthorized deletion of data" or other "denial of service or data" that are either inadvertent or intentional in nature.[18] Availability has alternatively been understood as "Ensuring timely and reliable access to and use of information."[19] Parker defines availability simply as the "Usability of information for a purpose."[20] Dardick's synthetic CFA model understands the joint concept of "Availability/Timeliness" as the "the degree to which the facts and analysis are available and relevant (valid and verifiable at a specific time)."[21] *NIST SP 800-33* contends that availability is commonly an organization's most important security goal.[22] Placing such a high priority on availability is reasonable, for example, in the case of an ad-vanced neuroprosthesis that provides its user with real-time sense data or support in cognitive processes such as memory or volition, where the loss of availability of the device and its data at a critical moment could result in in-jury or death.

The goals of confidentiality, integrity, and availability display a number of mutual interdependencies; for example, if a system's integrity has been lost, its mechanisms for maintaining the confidentiality and availability of its data may no longer be functional or reliable.[23] Some definitions of availability com-bine two or more different InfoSec goals by stating that the objective is not only to ensure that data is always available to legitimate users for legitimate purposes but also to ensure that it is always *unavailable* to any person or pro-cess that is attempting to use the data (or the larger system) for unauthorized ends.[24] In the framework presented here for neuroprosthetic devices, ensur-ing availability does not involve preventing information from being accessed by unauthorized parties (or authorized parties who would attempt to use the

---

[18] *NIST SP 800-33* (2001), p. 2.

[19] 44 U.S.C., Sec. 3542, cited in *NIST SP 800-37* (2010), Rev. 1, p. B–2.

[20] Parker (2002), p. 124.

[21] Dardick (2010), p. 61.

[22] *NIST SP 800-33* (2001), p. 2.

[23] *NIST SP 800-33* (2001), p. 4.

[24] *NIST SP 800-33* (2001), p. 2.

information for unauthorized purposes); instead, InfoSec goals such maintaining the *possession* and *confidentiality* of information represent those objectives.

D. Possession

With regard to an advanced neuroprosthetic system, we can define possession as "holding and controlling the physical substrate or substrates in which information is embodied." This definition requires that in order to have possession of information, the user of a neuroprosthetic device must have *sole* possession. If two different parties own physical copies of some information, then it can be said that the information is *available* to both parties but that neither 'possesses' it, insofar as neither party, acting individually, has the ability to prevent the creation of additional physical copies of the information or the distribution of such copies to additional parties.

Within the framework of the classic CIA Triad, possession is not explicitly defined as a freestanding security goal; however, it can be understood implicitly as an aspect of confidentiality and, in some cases, a prerequisite for maintaining integrity and availability. Possession is explicitly delineated as an independent InfoSec goal in the expanded Parkerian Hexad, where it is defined as "Holding, controlling, and having the ability to use information."[25] Meanwhile, Dardick's synthetic CFA model summarizes the joint concept of "Possession/Control" as relating to the 'chain of custody' of information.[26]

E. Authenticity

In the context of an advanced neuroprosthetic system, we can define authenticity as "the quality of in fact deriving from the source or origin that is claimed or supposed to be the information's source or origin." For example, the human host and user of a pair of artificial eyes might reasonably assume that the visual sense data presented by the eyes represents an accurate depiction of the physical environment surrounding the host. If the artificial eyes are presenting the host with the visual experience that he is sitting in his office at work while in fact he has been kidnapped and is sitting in a laboratory in the headquarters of a rival company – with his artificial eyes having been hacked to provide him with a false impression of his surroundings – we could say that the information being provided by the artificial eyes is inauthentic.[27]

---

[25] Parker (2002), p. 125.

[26] Dardick (2010), p. 61.

[27] For the possibility that a device designed to receive raw data from the external environment could have that data replaced with other data transmitted from some external information sys-

On the other hand, if – as an alternative means of taking a 'vacation' – a neuroprosthetic device's host had purposefully paid a sensory engineer to provide her with the visual experience of lounging on a tropical beach while in fact she were lying on her couch at home, we might say that this experience was 'virtual' or 'fabricated,' but according to our definition it would not be 'inauthentic,' because the host knew that the source of her sense data was not an actual physical beach surrounding her.[28] In other words, the host would not be having an experience of lounging on a real beach that is inauthentic but rather an authentic experience of lounging on a virtual beach.

Note that according to our definition, in order for some information provided by a neuroprosthesis to be inauthentic it is not required that someone have *explicitly claimed* that this particular information is accurate or originates from a source that is not its actual source; it is enough for the device's host or user to *suppose* that the information is originating from some source which is, in fact, not the actual source of the data. If a particular neuroprosthesis were sold without any claim that it will provide accurate and authentic information, but its human host has utilized the device for some time and has always found it to present an accurate and authentic representation of the physical environment surrounding him or her, then the user might understandably come to assume that this will always be the case in the future. If the device were then hacked and began to present the user with a stream of sense data that was inaccurate and did not reflect physical reality, that information could well be described as 'inauthentic,' from the user's perspective.

Gray areas may arise especially when neuroprostheses are being purposefully used to immerse their users in fabricated virtual environments. In general, it is more difficult to describe the information presented by a device as 'inauthentic' if the user knows that the purpose of the device is to present a fabricated virtual experience; however, it is still possible. For example, imagine that all of a multinational company's employees use neuroprosthetic devices that create a shared virtual environment in which employees from around the world can interact. If a hacker were to manipulate the sense data provided to one particular employee so that he or she believed that a coworker had just made a statement within the virtual world which, in fact,

<hr>

tem, see Koops & Leenes, "Cheating with Implants: Implications of the Hidden Information Advantage of Bionic Ears and Eyes" (2012). Regarding the possibility of neuroprostheses being used to provide false data or information to their hosts or users, see also McGee, "Bioelectronics and Implanted Devices" (2008), p. 221.

[28] In a similar way, one might say that a novel that claims to be historically accurate but is full of errors and anachronisms is 'inauthentic,' while the same work – if explicitly marketed as a work of fantasy and creative fiction – could not be criticized for being 'inauthentic.'

the coworker had never made, the contents of that fabricated statement could be understood as inauthentic.

Within the classic CIA Triad, authenticity is not explicitly described as a security goal. It is included in the expanded Parkerian Hexad, where authenticity is defined as the "Validity, conformance, and genuineness of information."[29] Dardick's synthetic CFA model, meanwhile, summarizes the joint concept of "Authenticity/Original" as the "quality of being authentic or of established authority for truth and correctness – 'best evidence' (Validity)."[30]

F. Utility

With regard to an advanced neuroprosthetic system, we can define utility as **"the state of being well-suited to be employed for some particular purpose."** Information is not inherently useful or non-useful; it possesses utility only with regard to some specific purpose that has been chosen by a sapient agent, such as its human host or user. The same information could be useful to one person in one moment but not useful to a different person or in a different moment.

For example, an artificial eye might generate sense data that is of use to its human host in reading, working at a computer, cooking, navigating his or her environment, or carrying out countless other everyday activities but which is not useful (and may even be distracting and detrimental) if the user is attempting to meditate, sleep, or concentrate on some mental task. Moreover, the device itself ceases to generate information that is even potentially useful when the eyelids in front of it are closed.[31] Other kinds of advanced neuroprostheses might not generate any information that is immediately useful to their host but may generate vast quantities of biological and diagnostic data that is useful to the team of medical personnel or engineers who are monitoring and controlling a device in order to effectuate some particular outcomes.

The concept of utility is not explicitly incorporated into the classic CIA Triad – though information's potential utility could be understood, for example, as the reason *why* one wishes certain information to be 'available.' Dar-

---

[29] Parker (2002), p. 125.

[30] Dardick (2010), p. 61.

[31] Even with the eyelids closed, an artificial eye conveys very basic information about whether the external environment surrounding its host is pitch black, moderately illuminated, or brightly illuminated. In some cases it may be desirable to eliminate the eyelids' ability to close (either to blink or while the host is asleep) in order to allow images of the external environment to be recorded and stored or transmitted by the eyes, even if they are not immediately presented to or consciously experienced by the host himself or herself.

dick's synthetic CFA model summarizes the joint concept of "Utility/Relevance" as providing an answer to the question "Is it useful? / is it the right information?"[32] As part of his security hexad, Parker defines utility as the "Usefulness of information for a purpose"[33] – which in the case of an advanced neuroprosthesis could be a purpose defined by the device's human host, by medical personnel or 'experiential engineers' who maintain and control the device with the host's permission in order to produce particular effects for the host, or potentially by some agent that has installed the neuroprosthesis without its host's knowledge or permission and which is utilizing the device to advance its own objectives. The latter might be the case, for example, with a neuroprosthetic that is implanted in an infant at the request of its parents, in a comatose individual at the request of his or her guardian, or by a military agency into its personnel or corporation into its employees. In such cases, the questionable legality and ethicality of such operations is not being considered here, only the fact that regardless of by whom or for what purpose a neuroprosthesis has been implanted, the party who has implanted it will see the device's ongoing utility as an objective to be pursued and whose loss would compromise the device's information security.

G. Distinguishability

In the context of an advanced neuroprosthetic system, we can define distinguishability as **"the ability to differentiate the information to be secured from information possessing a different source or nature."**[34] It is relatively easy to distinguish the system and data whose information security one is seeking to ensure in the case of a desktop computer, laptop computer, or mobile device: such devices are discrete units that can be identified and physically separated from their environments. Moreover, it is relatively easy to identify what data is stored on such a device, whether it be stored on a magnetic hard drive, flash memory, ROM chip, or some other physical substrate. By knowing the boundaries of one's system and identifying the information that is to be protected, one can thus develop a clear InfoSec strategy. However, because of their close integration with the human mind and body's own systems for generating, receiving, storing, transmitting, and processing information, it can be difficult to determine: 1) which are the synthetic systems and neuroprosthetically derived information which the designer, manufacturer, and operator of a neuroprosthesis may possess the legal and ethical authority to control and manipulate (and for whose security they may bear both legal and ethical responsibility), and 2) which are the natural biological systems of the host's

---

[32] Dardick (2010), p. 61.

[33] Parker (2002), p. 125.

[34] See Gladden, "Information Security Concerns as a Catalyst for the Development of Implantable Cognitive Neuroprostheses" (2016).

body and informational content of the host's mind – which the operator of a neuroprosthetic device may have a legal and ethical responsibility to keep secure, without necessarily possessing a legal basis for controlling, manipulating, or even affecting those systems and sources of information.[35] If the information provided by a neuroprosthesis cannot be distinguished from information emanating from other sources, it likely becomes more difficult to ensure the information's security.

Consider a human being who has received artificial retinal implants that supply the visual sense data constituting 30% of the person's field of vision, while the remaining 70% of the sense data is provided by the person's natural biological retinal cells. If the person knows which 30% of her field of vision is being generated by her neuroprostheses, her InfoSec situation is qualitatively different from that of a person who knows that 30% of his field of vision is being provided by an artificial device but does not know *which portion* of his field of vision is 'synthetic' and which is 'natural.' Similarly, the user of a mnemocybernetic implant who is able to easily distinguish (e.g., through some ineffable inner sensation or awareness) the mnemonic content provided by the implant from the mnemonic content stored in his or her brain's natural memory systems faces a different InfoSec situation than someone who knows that he or she possesses a mnemonic implant but has no way of distinguishing memories stored in the implant from memories stored in the natural mechanisms of his or her brain – and different still is the situation of someone who does not even realize that he or she possesses a mnemonic implant and who is thus not even aware that he or she should be *attempting* to distinguish between those memories that are natural and those that are neuroprosthetically generated.

H. Rejectability

With regard to an advanced neuroprosthetic system, we can define rejectability as **"the ability to exclude particular information from one's conscious awareness on the basis of its source, nature, or other characteristics."**[36] It is important to ensure that information is available whenever its user wishes to access it. However, in the case of a neuroprosthetic device it is at least as important to ensure that information is not involuntarily forced into the mind of its host or user when he or she *does not* wish to access it.

The ability of advanced neuroprostheses to forcibly inject experiences – whether sense data, memories, emotions, or other mental phenomena – into

---

[35] With respect to the complex questions that arise regarding who bears moral, legal, and financial responsibility for activities involving implanted ICT devices, see Roosendaal, "Carrying Implants and Carrying Risks; Human ICT Implants and Liability" (2012).

[36] See Gladden, "Information Security Concerns as a Catalyst for the Development of Implantable Cognitive Neuroprostheses" (2016).

the conscious awareness of their host or user makes it essential that such devices have safeguards to guarantee that their hosts and users are not subject to sensory overload, brainwashing, or other kinds of psychological or emotional assault. This becomes particularly important if, for example, a neuroprosthetic device has been implanted in a child, an individual suffering motor impairments, or other persons who may not be able to actively adjust or disable the device or express their lack of consent to the experience. Cognitional security involves not only being able to bring desired information into one's mind for use but also to keep it out of one's mind, when desired.

I. Autonomy

In the context of an advanced neuroprosthetic system, we can define autonomy as **"the state of a subject that consciously experiences its own use of information and which possesses and exercises agency in generating information."**[37] This is clearly not an attribute that applies to information stored in a traditional system like the hard drive of a desktop computer. In that case, neither the information itself nor the computer system containing the information possesses a subjective experience of the information, and if the computer can be said to exercise 'agency' in generating information, it is only in a limited sense (at least, in comparison to human beings), insofar as a conventional computer does not possess its own desires, beliefs, volitions, or conscience.[38]

In the case of a human being implanted with an advanced neuroprosthesis, the information contained within the device does not, in itself, possess autonomy. However, unless he or she is in state (such as that of a coma) that deprives him or her of the ability to consciously experience information and

---

[37] See Gladden, "Information Security Concerns as a Catalyst for the Development of Implantable Cognitive Neuroprostheses" (2016).

[38] 'Weak' notions of agency define an agent as any entity that displays the externally observable characteristics of autonomy, reactivity, proactivity, and an ability for social interaction; 'strong' notions of agency insist that an agent also possess internal mental phenomena such as beliefs and desires (which, when joined, can constitute intentions). For these definitions of agency, see Wooldridge & Jennings, "Intelligent agents: Theory and practice" (1995), and Lind, "Issues in agent-oriented software engineering" (2001). For more on the relationship of beliefs, desires, and intentions, see Calverley, "Imagining a non-biological machine as a legal person" (2008). Regarding the extent to which it is possible for technological devices – whether a conventional desktop computer or a far more sophisticated construct such as a social robot or artificial general intelligence (AGI) – to possess and demonstrate agency and autonomy and the forms that these traits can take, see, e.g., Coeckelbergh, "From Killer Machines to Doctrines and Swarms, or Why Ethics of Military Robotics Is Not (Necessarily) About Robots" (2011); Calverley (2008); Hellström, "On the Moral Responsibility of Military Robots" (2013); Kuflik, "Computers in Control: Rational Transfer of Authority or Irresponsible Abdication of Autonomy?" (1999); Stahl, "Responsible Computers? A Case for Ascribing Quasi-Responsibility to Computers Independent of Personhood or Agency" (2006); and Friedenberg, *Artificial Psychology: The Quest for What It Means to Be Human* (2008).

to utilize agency in generating it, the potential human host of a neuroprosthetic device *does* possess informational autonomy in the sense defined above, and the integration of the neuroprosthesis into the host's neural circuitry to create a new host-device system should not be allowed to impair or destroy the host's informational autonomy. In this sense, we can say that preserving the informational autonomy of the host-device system is an important goal of information security. Autonomy is thus the epitome of a new kind of InfoSec goal and attribute that has not been relevant for the information security of traditional computerized information systems but which becomes relevant – and, indeed, assumes paramount importance – in the case of information stored within an advanced neuroprosthesis and the larger system that it forms with its human host.

When working to ensure the security of information contained within a hard drive, the hard drive does not possess its own rights about which we must be concerned. Similarly, a computer running the most sophisticated sorts of artificially intelligent software available today may demonstrate a limited form of agency, but such a platform is not a moral agent that is capable of possessing its own conscience or conscious awareness, nor is it (like infants and at least some animals) the sort of 'moral patient' about whose welfare human beings must be concerned even though it is not in itself a moral agent.[39] Thus it may be appropriate for an organization to routinely destroy hard drives or entire disused computers as part of an overall strategy for keeping secure the information contained within them. However, in the case of a neuroprosthetic device, maintaining the biological and psychological welfare of the being into whose organism and mind the device is integrated is typically the greatest priority, and any efforts at securing the neuroprosthetic device and information contained within it must not be allowed to impair the well-being of the device's human host.

Note that there is not simply a danger that a device's built-in InfoSec mechanisms might harm its host; it is also possible that the device's mere presence might damage its host's mind and the information contained within it. Especially in the case of neuroprostheses that affect their host's processes

---

[39] Regarding the distinctions between legal persons, moral subjects, and moral patients – particularly in the context of comparing human and artificial agents – see, e.g., Wallach & Allen, *Moral machines: Teaching robots right from wrong* (2008); Gunkel, *The Machine Question: Critical Perspectives on AI, Robots, and Ethics* (2012); Sandberg, "Ethics of brain emulations" (2014); and Rowlands, *Can Animals Be Moral?* (2012).

of memory,[40] volition, metavolition,[41] emotion,[42] or conscious awareness,[43] there is a possibility that the device might negatively impact the host's possession and exercise of the autonomy, moral agency, consciousness, and conscience that are among the defining traits of human beings.

It is possible to conceive of an invasive neuroprosthetic device which, for example, replaces sections of its host's brain in a way that destroys the host's conscious awareness while replacing it with an AI-driven artificial agency contained in the device.[44] Such concerns regarding authenticity and personal identity have already been expressed regarding neural implants used for deep brain stimulation to treat conditions such as Parkinson's disease.[45] We can summarize these concerns and this security goal by stating that a neuroprosthetic device should support rather than impair the autonomy of its human host and user.

III. Prioritizing the information security goals

Parker notes that however many security attributes one might define, they should be placed in some logical order (such as their order of importance) that adds an additional level of meaning to the list of attributes.[46] If we were

---

[40] For recent efforts at developing mnemoprosthetic technologies for mice that hint at the possibility of eventually developing similar technologies for human beings, see Han et al., "Selective Erasure of a Fear Memory" (2009), and Ramirez et al., "Creating a False Memory in the Hippocampus" (2013).

[41] See, for example, Negoescu, "Conscience and Consciousness in Biomedical Engineering Science and Practice" (2009), and Gladden, "Enterprise Architecture for Neurocybernetically Augmented Organizational Systems: The Impact of Posthuman Neuroprosthetics on the Creation of Strategic, Structural, Functional, Technological, and Sociocultural Alignment" (2016).

[42] For the possibility of developing emotional neuroprostheses, see Soussou & Berger, "Cognitive and Emotional Neuroprostheses" (2008); Hatfield et al., "Brain Processes and Neurofeedback for Performance Enhancement of Precision Motor Behavior" (2009); Kraemer, "Me, Myself and My Brain Implant: Deep Brain Stimulation Raises Questions of Personal Authenticity and Alienation" (2011); McGee, "Bioelectronics and Implanted Devices" (2008), p. 217; and Fairclough, "Physiological Computing: Interfacing with the Human Nervous System" (2010).

[43] For the possibility of neuroprosthetic devices relating to sleep, see Claussen & Hofmann, "Sleep, Neuroengineering and Dynamics" (2012), and Kourany, "Human enhancement: Making the debate more Productive" (2013), pp. 992-93.

[44] For a discussion of such possibilities, see Gladden, *Neuroprosthetic Supersystems Architecture* (2017), pp. 133-34.

[45] For the effects of existing kinds of neuroprosthetic devices on the agency (and perceptions of agency) of their human hosts, see Kraemer (2011) and Van den Berg, "Pieces of Me: On Identity and Information and Communications Technology Implants" (2012).

[46] Parker (2010), p. 17.

to arrange our nine InfoSec goals in order of importance as seen from the perspective of a generic neuroprosthetic host-device system (and in particular, a host's conscious awareness), a ranking that appears reasonable would be:

- **Autonomy** of the host-device system, insofar as a neuroprosthesis that impairs its host's autonomy, agency, conscience, and conscious awareness may actually destroy the host's most fundamental ability to experience and use information, thereby rendering all of the other security attributes irrelevant.

- **Rejectability**, as the ability of a hostmind to *block out* a stream of information that is causing ongoing pain, sensory overload, or physical or psychological trauma is arguably more important than the mind's ability to *access* information that is beneficial and useful.

- **Integrity**, which, if lost, would likely diminish or destroy the utility and authenticity of information contained in sense data and memories and render possession of that information of little value.

- **Utility**, as there is little need to ensure, for example, the availability or possession of information if it is ultimately of no use to the neuroprosthetic device's user.

- **Availability**, which may be crucial for information provided by some neuroprostheses (for example, sense data should be provided by an artificial eye in real time, in order to synchronize with data provided by other sensory organs and allow real-time motor control) but less important for information provided by others (for example, a delay in retrieving certain kinds of long-term memories stored in a memory implant may be permissible).

- **Confidentiality**, insofar as a neuroprosthetic device may potentially allow outside agents to access the contents of its host's volitions, memories, emotions, and other intimate mental processes whose contents the host would very much like to keep private.

- **Authenticity**, insofar as information that is 'false' or 'inauthentic' (such as fabricated sense data that intentionally misleads a device's host into believing that he is walking through a forest while in fact he is lying on a bed in a hospital) may still be of great value to the device's host and operator as long as it is useful and available.

- **Possession**, as the permanent holding and control of some information provided by neuroprostheses (such as long-term memories) may be important, but the ability to store long-term and to control other kinds of information (such as a complete permanent record of all the visual information that one has experienced through one's retinal prostheses) that are generally experienced only instantaneously by the mind is something that natural biological human beings do not currently enjoy and may not require neuroprostheses to grant.

- **Distinguishability,** insofar as it may not be important to distinguish neuroprosthetically supplied from naturally supplied information, as long as the neuroprosthetically supplied information possesses all of the other InfoSec attributes. However, distinguishability becomes an important tool that is useful for pursuing information security in cases where other attributes are lacking and specific vulnerabilities, threats, or risks need to be addressed.

Note that if one accepts such an ordering, two of the three new InfoSec goals that we have formulated for advanced neuroprostheses turn out to be more important than any of the goals traditionally defined in the CIA Triad or Parkerian Hexad. This highlights the peril of assuming that security goals that were developed with previous standalone computing devices (like conventional desktop or laptop computers) in mind will provide an adequate basis for ensuring information security for new kinds of neuroprosthetic devices that are intimately interconnected with the biological and mental processes of a human user. This underscores the need to develop new and more robust cognitional security frameworks for such brain-machine interfaces.

Although the ranking of InfoSec attributes proposed above appears reasonable as a generic approach, many alternative rankings are possible. While Parker seems to suggest that there may be a single most logical way of ordering security attributes,[47] other experts note that different organizations will prioritize such attributes in different ways,[48] based on each organization's unique mission and the role that information and information technology play within it. In the case of an advanced neuroprosthetic system, any prioritized ordering of security attributes includes many implicit value judgments about the relative importance of various objectives, and different individual hosts or users of neuroprostheses might rank the attributes in quite different ways.

For example, for device hosts who are powerful political figures or business leaders, ensuring the confidentiality of information contained within their minds might be the ultimate priority, insofar as that information may include classified national security plans that must not be allowed to fall into the hands of hostile states or trade secrets that could be exploited by competing firms; moreover, such a person's mind may contain information (like long-term memories dating back to childhood) which, if acquired by unauthorized parties, could provide a basis for blackmail, extortion, or other illicit manipulation. On the other hand, other users might be willing to accept a

---

[47] Parker (2010), p. 17.
[48] *NIST SP 800-33* (2001), p. 2.

loss of confidentiality, if in return their neuroprostheses would grant them new sensorimotor or cognitive capacities, allow them to interact socially in new ways, or provide them with other advantages that outweigh the loss of confidentiality. Indeed, there is even reason to believe that over time, some human beings may come to embrace the use of neuroprostheses that allow members of a community to mutually experience one another's thoughts – thereby purposefully reducing the confidentiality of information contained within such hosts' minds in order to forge new kinds of political dialogue and social relations; in such cases, a loss of confidentiality could be experienced as something 'liberating' that advances openness and honesty rather than something frightening and oppressive.[49]

Similarly, some users may give paramount value to the authenticity of the information being conveyed by their neuroprosthetic device. Such users might prefer to have an artificial eye which, for example, provides them with a stream of low-resolution visual sense data that is not particularly useful but which they know is 'authentic' (i.e., it accurately reflects the objective physical reality of the environment surrounding them) rather than possess an artificial eye that provides flawless high-resolution video but which can easily be hacked by unauthorized parties – so that a device's host never knows whether the world that he or she is seeing actually exists or whether it is a false or virtual environment that he or she is experiencing as a result of fabricated sense data that is being fed to the neuroprosthetic device by a malicious hacker.

When prioritizing the InfoSec goals for a particular advanced neuroprosthetic system, the system's designer should thus take into account factors such as:

- The **market segment(s) of potential users** at whom the device is being targeted.
- The unique information security **needs and concerns** manifested by those groups.

---

[49] The prospect of creating 'hive minds' and neuroprosthetically facilitated collective intelligences is investigated, e.g., in McIntosh, "The Transhuman Security Dilemma" (2010); Roden, *Posthuman Life: Philosophy at the Edge of the Human* (2014), p. 39; and Gladden, "Utopias and Dystopias as Cybernetic Information Systems: Envisioning the Posthuman Neuropolity" (2015). For classifications of different kinds of potential hive minds, see Chapter 2, "Hive Mind," in Kelly, *Out of Control: The New Biology of Machines, Social Systems and the Economic World* (1994); Kelly, "A Taxonomy of Minds" (2007); Kelly, "The Landscape of Possible Intelligences" (2008); Yonck, "Toward a standard metric of machine intelligence" (2012); and Yampolskiy, "The Universe of Minds" (2014). For critical perspectives on the notion of hive minds, see, e.g., Maguire & McGee, "Implantable brain chips? Time for debate" (1999); Bendle, "Teleportation, cyborgs and the posthuman ideology" (2002); and Heylighen, "The Global Brain as a New Utopia" (2002).

- The ways in which the InfoSec characteristics of the neuroprosthetic device itself, its physical maintenance services, its software updates, and other related products and services offered by the manufacturer will **integrate with the existing information security systems**, services, and priorities maintained by institutions (such as employers, schools, health care providers, or government agencies) that already bear responsibility for ensuring those users' information security.

Moreover, in order for the potential host of a neuroprosthetic device to choose the device that is best for him or her and to provide informed consent for its implantation, the relative prioritization of information security goals that have been incorporated into the device's design and functioning should be disclosed to the potential host in the relevant marketing materials and pre-implantation counseling.

IV. Understanding the security goals at three levels

The human host of an advanced neuroprosthetic device intertwines his or her personal information security with that of the device on different levels, each of which has distinct InfoSec challenges and characteristics that must be taken into consideration.[50] We can consider such a neuroprosthetically enabled human being on at least three different levels:[51]

---

[50] Note that while considering the human host of a neuroprosthetic device separately as a sapient mind, embodied biological organism, and social and economic actor is useful for ensuring that one does not overlook any of the InfoSec issues that become especially apparent when considering the human host in one of those capacities, in reality these three roles are deeply interrelated, if not wholly inextricable from one another. In future posthumanized contexts in which very sophisticated neuroprostheses have been deployed, it may sometimes be clear, e.g., that a particular human being has become 'infected' by a particular *idée fixe* that occupies all of his or her thoughts or has become wrapped up in a relationship (such as one of love, loyalty, or hatred) that consumes all of the person's energy and attention – but it may be unclear whether the source of the phenomenon, the vector that introduced it into the person's mental life – was a biological vector (such as a biological virus or biochemical agent that has affected the person neurologically or physiologically), an electronic vector (such as glitches that occurred in the gathering of sense data or storage of memories by a neuroprosthesis or malware that has infected the synthetic components of the host-device system), or a social vector (such as acts of inspiration, persuasion, seduction, blackmail, or myth-building performed by other intelligent agents and directed at the person). For a use of actor-network theory (ANT) to explore the ways in which, e.g., a single 'idea' might manifest itself through diverse biological, mental, technological, and social phenomena and the complexities involved with untangling biological and technological symbioses and power relations within such a posthumanized context, see Kowalewska, "Symbionts and Parasites – Digital Ecosystems" (2015).

[51] See Gladden, "Neural Implants as Gateways to Digital-Physical Ecosystems and Posthuman

1. The human being as a **sapient metavolitional agent**, a unitary mind that possesses its own conscious awareness, memory, volition, and conscience (or 'metavolitionality'[52]).

2. The human being as an **embodied embedded organism** that inhabits and can sense and manipulate a particular environment through the use of its body.

3. The human being as a **social and economic actor** who interacts with others to form social relationships and to produce, exchange, and consume goods and services.[53]

At each of these three levels, a neuroprosthetic device integrates with its host's own natural capacities to create a host-device system whose unique characteristics may create powerful new tools that can assist with ensuring the system's information security, serious new vulnerabilities that undermine the system's information security, or both. Below we consider some new capacities and limitations that a neuroprosthesis can provide its human host at each of the three levels and describe the impact that these new characteristics can have on pursuit of the nine InfoSec goals.

A. Functional vs. information security impacts

Note that there is no direct correlation between a neuroprosthetic device having an *overall functional impact* on its host that is considered positive or negative and its having a more specific *information security impact* that is considered positive or negative. Some new neuroprosthetically facilitated characteristics that might be considered *beneficial* from a host's perspective (due to the new functional capacities that they provide) may be considered harmful and disadvantageous from the perspective of the InfoSec professionals who are charged with ensuring the host's information security, insofar as the characteristics create egregious new vulnerabilities. Conversely, some aspect of a neuroprosthetic device that is generally considered undesirable from the perspective of its host (because it limits or constraints the host in some way) might be considered advantageous from an InfoSec perspective, insofar as it provides a new layer of defense that protects information contained

---

Socioeconomic Interaction" (2015), on which the reminder of this chapter draws heavily.

[52] See Calverley (2008) for an explanation of the relationship of second-order volitions to conscience and Gladden (2017), pp. 231-33, for use of the word 'metavolitional' in this context regarding neuroprosthetic devices.

[53] The financial and economic aspect of a neuroprosthetic device's impact is important, insofar as financial considerations influence the kind and degree of security measures that can be implemented by individual or institutional users of neuroprostheses, and efforts to compromise information security and illicitly acquire information often have a financial component (e.g., as part of a planned scheme for blackmail, corporate espionage, or sale of the information).

within the host's biological systems or mental processes from access by un-authorized parties.

B. Impacts on a host vs. impacts on a user

Note also that the impacts that a particular neuroprosthetic device has on the information security of its human *host* may differ significantly (and even be diametrically opposed to) the impacts that it has on the information security of its operator or *user*, if the host and user are different persons. In cases where the host and user are different individuals, a cognitional security framework should be applied separately to the device's host and its user and due attention should be paid to the impacts result for each person.

C. The human host as sapient metavolitional agent

1. Functional capacities created by a neuroprosthetic device

Below we describe some of the new functional capacities that a neuropros-thetic device can provide its host in his or her role as a sapient metavolitional agent and the potential impact that these capacities might have on the information security of the host-device system.[54]

a. Enhanced memory, skills, and knowledge stored within the mind (engrams)

Building on current experimental technologies that are being tested in mice, future neuroprostheses may offer human hosts the ability to create, al-ter, or weaken memories that are stored in their brains' natural memory systems in the form of engrams.[55] Such technologies could potentially be used not only to affect a user's declarative knowledge but also to enhance motor skills or reduce learned fears.

Tremendous technological challenges would need to be overcome in order to someday develop a neuroprosthetic device that allows for the precise 'ed-iting' of extant human memories or creation of complex new memories within the brain's naturally existing memory systems. Indeed, the exact struc-tures and processes used by the brain to encode, store, and retrieve long-term memories are still shrouded in mystery, and researchers have proposed di-vergent theories to account for the way in which the brain stores engrams.[56] If, for example, holographic models such as the Holonomic Brain Theory are

---

[54] See Gladden, "Neural Implants as Gateways to Digital-Physical Ecosystems and Posthuman Socioeconomic Interaction" (2016).

[55] See Han et al. (2009); Ramirez et al. (2013); McGee (2008); and Warwick, "The Cyborg Revolu-tion" (2014), p. 267.

[56] For a discussion of such unresolved scientific questions, see, e.g., Dudai, "The Neurobiology of Consolidations, Or, How Stable Is the Engram?" (2004).

correct, then any efforts to make precise adjustments to existing memories by manipulating neurons in a particular portion of the brain may prove futile, as each memory may be stored holographically across the brain's entire neural network (or at least, across a large portion of it).[57] Although researchers have succeeded in understanding many of the large-scale synaptic structures and basic electrochemical functioning of neural synapses – and are making rapid progress at developing artificial neurons that can replicate key elements of this observed synaptic functioning – there is still considerable debate about the extent to which these simple, large-scale synaptic structures and activities within the brain are responsible for the creation, storage, and recall of long-term memories.[58] The Holonomic Brain Theory, for example, proposes that much more sophisticated and difficult-to-observe interactions between neurons (such as those within the 'synaptodendritic web'[59]) – may play essential roles in the memory process that we have barely begun to comprehend. The development of a neuroprosthetic device that can successfully integrate with the brain's neural circuitry in order to support, expand, control, or replace natural mechanisms for creating, storing, and retrieving complex engrams is not expected to occur soon, and – depending on which theories of the brain's memory processes prove correct – it may not even be theoretically possible at all.

However, if one assumes that such a technology can be developed, it is clear that it would have major implications for the security of information held within the long-term memory of its human host. If a device were to allow external agents to access and copy a mind's engrams, this would imperil that information's **confidentiality** as well as the host's **possession** of it. The ability to edit or delete existing engrams would threaten the information's **integrity, utility, availability**, and **authenticity** for the host. If the device were able to retrieve particular memories and present them to the host or user's conscious awareness against his or her will, it would undermine the **rejectability** of that information. If the device provided an external agent the wholesale ability to delete, replace, or manipulate its host's memories, this could potentially reduce the host's **autonomy** by eliminating his or her own ability to exercise agency in generating mnemonic contents. If the device were integrated seamlessly into the brain's natural mnemonic systems, from its host's perspective the

---

[57] For a discussion of holographic models of the brain, see, e.g., Longuet-Higgins, "Holographic Model of Temporal Recall" (1968); Westlake, "The possibilities of neural holographic processes within the brain" (1970); Pribram, "Prolegomenon for a Holonomic Brain Theory" (1990); and Pribram & Meade, "Conscious Awareness: Processing in the Synaptodendritic Web" (1999). An overview of conventional contemporary models of long-term memory is found in Rutherford et al., "Long-Term Memory: Encoding to Retrieval" (2012).

[58] See, e.g., Dudai (2004).

[59] See Pribram & Meade, "Conscious Awareness: Processing in the Synaptodendritic Web – The Correlation of Neuron Density with Brain Size" (1999).

(potentially inauthentic) memories generated by the neuroprosthesis might lack **distinguishability** from natural memories that were generated by some actual experience in the host's past.

b. Enhanced creativity

A neuroprosthesis may be able to enhance a mind's powers of imagination and creativity by facilitating processes that contribute to creativity, such as stimulating mental associations between unrelated items. Anecdotal increases in creativity have been reported to occur after the use of neuroprostheses for deep brain stimulation.[60]

If such a device were able to force new thoughts into its host's mind against his or her will, that information would lack **rejectability**, and the device might undermine **autonomy** by interfering with or overriding the host's ability to generate his or her own creative thoughts. Moreover, by forcing unwanted and distracting memories into the host's conscious awareness, this could interfere with the host's efforts to access *other* information contained within his or her memory, thereby reducing the **availability**, **utility**, and potentially **integrity** of the latter information. If the host could never be sure whether new ideas had been generated by his or her own imagination or by the device, those ideas would lack **distinguishability**. The ability of outside agents to access ideas generated by the device would undermine that information's **confidentiality** and the host's **possession** of it.

c. Enhanced emotion

A neuroprosthetic device might provide its host with more desirable emotional dynamics and behavior.[61] Effects on emotion have already been seen, for example, with devices used for deep brain stimulation.[62]

If such a device were to allow external agents to detect its host's internal emotional states, the device would be undermining the host's **possession** and the **confidentiality** of that information. If the device could force emotional content into the host's conscious awareness, that information would lack **rejectability** and could undermine the host's **autonomy**. Insofar as such involuntary emotional dynamics distort or render impossible the host's ability to efficiently access and use other information, the **availability**, **utility**, and perhaps **integrity** of such information would suffer.

On the other hand, in the case of a host whose previous severe emotional disturbances had made it difficult or impossible for the person to calmly and

---

[60] See Cosgrove, "Session 6: Neuroscience, brain, and behavior V: Deep brain stimulation" (2004), and Gasson, "Human ICT Implants: From Restorative Application to Human Enhancement" (2012).

[61] McGee (2008), p. 217.

[62] See Kraemer (2011).

efficiently access and utilize information contained within his or her memory or provided by the external environment, the use of such a device could potentially *enhance* the **availability** and **utility** of such information. If a person is prone to fits of uncontrollable anger, jealousy, or pride during which he or she lashes out and reveals his or her harshest critiques of others or other personal secrets, the use of a neuroprosthetic device to limit such frustrations and outbursts could aid its host in maintaining the **confidentiality** and **possession** of information which, in moments of greater rationality, the person would admit that he or she has no desire to reveal. By giving the host greater control over his or her emotions, such a device could enhance the person's own agency (and thus informational **autonomy**) and increase the **rejectability** of unwanted thoughts and feelings that the host was previously unable to block out of his or her mind.

d. Enhanced conscious awareness

Research is being undertaken to develop neuroprostheses that would allow the human mind to, for example, extend its periods of attentiveness and limit the need for periodic reductions in consciousness (i.e., sleep).[63]

By enhancing the mind's attentiveness and ability to spend extended periods of time focused on accessing and processing information, such a device could indirectly enhance the **availability** and **utility** of information for its host. Enhancing and extending the host's conscious awareness could also temporally expand (if not otherwise qualitatively change) the host's ability to exercise agency and **autonomy** in the accessing and use of information. On the other hand, if a device is capable of forcibly compelling the host to focus his or her conscious awareness on a particular piece of information, that would limit the **rejectability** of that information and weaken the host's **autonomy**. Moreover, if the device could be misused to *reduce* the host's conscious awareness, this would impair his or her ability to subjectively experience information and thus reduce the host's **autonomy**.

e. Enhanced conscience

If a 'volition' is understood as a belief about the outcome of some action and a desire for that outcome, then one's conscience can be understood as one's set of second-order volitions or 'metavolitions' – desires about the kinds of volitions that one wishes to possess.[64] Insofar as a neuroprosthetic device enhances processes of memory and emotion that allow for the development of one's conscience, the device may enhance one's ability to develop, discern, and follow one's conscience.

---

[63] Kourany (2013), pp. 992-93.
[64] See Calverley (2008), pp. 528-34, and Gladden (2017), pp. 231-33.

A neuroprosthetic device that is capable of altering its host's most funda-mental desires and assessment of what is 'right' and 'wrong' would have major implications for information security – most noticeably in either strengthen-ing or undermining the host's informational **autonomy** and potentially impair-ing the **integrity** and **availability** of the information that would be conveyed by the host's unaugmented conscience in the absence of such a device. By affect-ing the host's metavolitions, such a device would over time alter the **rejecta-bility** and availability of information contained in the host's first-order voli-tions. If such a device provides the host with metavolitions that are deter-mined by (and thus known to) some external agent, the host's previous metavolitions (over which he or she presumably exercised **possession** and sole control) would be replaced by new metavolitions lacking **confidentiality** and sole **possession** by the host.

2. Functional impairments created by a neuroprosthetic device

Below we describe some of the functional impairments that a neuropros-thetic device might create for its host at the level of his or her internal mental processes and the impact that these impairments might have on the infor-mation security of the host-device system.[65]

a. Loss of agency

A neuroprosthesis may damage the brain or disrupt its activity in a way that reduces or eliminates the ability of its human host to possess and exer-cise agency.[66] Moreover, the knowledge that this can occur may lead hosts to doubt whether their volitions are really 'their own' – an effect that has been seen with neuroprostheses used for deep brain stimulation.[67]

A neuroprosthetic device that produces a general loss of agency would clearly have a negative impact on its host's informational **autonomy** by reduc-ing the host's ability to possess and exercise agency in generating infor-mation. It could also indirectly reduce the **rejectability** of unwanted infor-mation and the **availability** of desired information.

b. Loss of conscious awareness

A neuroprosthesis may diminish the quality or extent of its host's con-scious awareness – for example, by inducing daydreaming or increasing the required amount of sleep. A neuroprosthesis could potentially even destroy

[65] See Gladden, "Neural Implants as Gateways to Digital-Physical Ecosystems and Posthuman Socioeconomic Interaction" (2016).

[66] McGee (2008), p. 217.

[67] See Kraemer (2011).

its host's capacity for conscious awareness (e.g., by inducing a coma) but without causing the death of his or her biological organism.[68]

A neuroprosthetic device that produces a loss of conscious awareness on the part of its host would have a negative impact on its host's informational **autonomy** similar to that produced by a loss of agency; if a host's ability to subjectively experience information is completely destroyed, then other attributes such as the **integrity**, **utility**, and **availability** of information for that host would become largely irrelevant, as there would no longer *be* a 'host' to whom the information could be presented.

c. Dependency of internal cognitive processes on external systems

Although the portion of a neuroprosthetic device that directly interfaces with its host's neural circuitry is typically implanted in its host's body, it is possible that internal processing, memory, and power constraints may force the device to regularly offload some information to an external system for processing (e.g., through a wireless data link) or to receive instructions from the external system. In this way, the 'internal' cognitive processes of the device's host may no longer be taking place solely within the relatively easily protected space of the host's brain and body but within an array of physically disjoint systems that communicate through channels that may be subject to accidental disruption or intentional manipulation.

The restructuring of the host's cognitive processes in such a way increases the possibility of a loss of **autonomy** and a reduction in the **integrity**, **availability**, **confidentiality**, **authenticity**, and **possession** of information contained in those cognitive processes. On the other hand, use of external systems to support or create a 'backup copy' of the host's internal cognitive processes could potentially also aid in the diagnosis and treatment of cognitive disorders, increased efficiency and power for the mind's cognitive processing, and the restoration of information that otherwise would have been lost to or by the brain's internal cognitive processes – all of which might contribute to an increase in **autonomy** and the **integrity**, **availability**, and **utility** of information.

d. Inability to distinguish a real from a virtual ongoing experience

If a neuroprosthetic device alters or replaces its host's sensory perceptions, it may make it impossible for the user to know which (if any) of the sense data that he or she is experiencing corresponds to some actual element of the primary external physical environment and which is 'virtual' or simply 'false.'[69]

---

[68] See Gladden (2017), p. 87.

[69] The term 'primary physical world' may be used to refer to what is commonly described as the 'real' physical world, whose contents possess an objective, independent existence; that world can

Such a neuroprosthetic device would certainly produce a loss of **distinguishability** in the sensory information experienced by its host and would open the door to external manipulation that could reduce the **availability** of accurate information that the host was blocked from seeing and the **authenticity** and **utility** of the information that was instead received by the host.

e. Inability to distinguish true from false memories

If a neuroprosthetic device is able to create, alter, or destroy engrams within its host's brain, it may be impossible for a host to know which of his or her apparent memories are 'true' and which are 'false' (i.e., distorted or purposefully fabricated).[70]

This kind of neuroprosthetic device would produce a loss of **distinguishability** in the mnemonic information experienced by its host and could facilitate external manipulation that would reduce the **availability** of accurate mnemonic information that the host was blocked from recalling and the **authenticity** and **utility** of the information that was instead recalled by the host. It could also impair the host's **autonomy**, insofar as he or she may end up exercising agency and making decisions based on memories that are not actually his or her own.

f. Other psychological side-effects

A host's brain may undergo potentially harmful and unpredictable structural and behavioral changes as it adapts to the presence, capacities, and activities of an advanced neuroprosthetic device.[71] These effects may even include new kinds of neuroses, psychoses, and other disorders unique to hosts or users of advanced neuroprostheses.

---

be contrasted with 'secondary physical worlds,' or virtual worlds whose contents are determined by the computational processes of a computerized virtual reality system. The contents of such virtual worlds are arbitrary, insofar as they are not constrained by the organization of the primary physical world and can be dramatically altered at will by a virtual world's human designer or world-management algorithms. Even secondary physical worlds are still 'real' and 'physical,' though, insofar as the structure of their contents is maintained within real physical objects (e.g., the hard drives or ROM chips of a VR computer system) and are experienced by their inhabitants through the mediation of real physical stimuli (such as electrons or chemical neurotransmitters used to stimulate neurons in a host's sensory system or brain). Regarding such distinctions, see Gladden (2017), pp. 128-30. For the possibility that a device designed to receive raw data from an external environment could have that data replaced with other data transmitted from some external information system, see Koops & Leenes (2012). Regarding the possibility of neuroprostheses being used to provide false data or information to their hosts or users, see also McGee (2008), p. 221.

[70] See Ramirez et al. (2013) for experimental technologies being tested in mice that have the potential to allow basic editing of memories.

[71] See McGee (2008), pp. 215-16, and Koops & Leenes (2012), pp. 125, 130.

Depending on their nature and severity, such changes could negatively impact hosts' **autonomy** and the **rejectability, integrity, utility, availability, authenticity,** and **distinguishability** of information experienced by the hosts or users, as well as potentially leading hosts to involuntarily disclose information in a way that damages its **confidentiality** and the hosts' **possession** of it. On the other hand, it is also possible that the structural and behavioral changes occurring in a host's brain as the result of using an advanced neuroprosthesis might have salutary effects that increase the host's **autonomy**, enhance the **integrity**, **availability**, and **utility** of information, and strengthen the host's ability to maintain the **confidentiality** and **possession** of that information. The designers of neuroprosthetic devices will need to conduct careful monitoring and testing to identify the short- and long-term effects of the devices' use and discover potentially unexpected side-effects that may have an impact on information security.

D. The host as embodied embedded organism

1. Functional capacities created by a neuroprosthetic device

Below we describe some of the new functional capacities that a neuroprosthetic device can provide its human host or user in his or her role as an embodied embedded organism and the potential impact that these capacities might have on the information security of the host-device system.[72]

a. Sensory enhancement

A neuroprosthetic device may allow its host to sense his or her physical or virtual environment in new ways, either by acquiring new kinds of raw sense data or new modes or abilities for processing, manipulating, and interpreting sense data.[73]

The **availability, integrity, utility,** and **authenticity** of information provided by such devices depend not only on their quality and technical specifications but also on securing the devices from external manipulation. If the sense data that is being gathered by a device and transmitted to its user's mind can be intercepted by external agents, the **confidentiality** and **possession** of that information is undermined. Such devices also raise questions of **rejectability** if a user cannot block out the information that they provide. The extent to which information provided by a neuroprosthetic device displays **distinguishability** from its user's other natural sensory input may depend not only on the device's technical capacities and limitations but also on explicit design decisions made by the device's producer about the ways in which information should

---

[72] See Gladden, "Neural Implants as Gateways to Digital-Physical Ecosystems and Posthuman Socioeconomic Interaction" (2016).

[73] See Warwick (2014), p. 267; McGee (2008), p. 214; and Koops & Leenes (2012), pp. 120, 126.

be presented. Insofar as such devices might expand the capacities of their users to consciously experience sense data and make decisions on the basis of it, such devices could potentially enhance their users' agency and **autonomy**.

b. Motor enhancement

A neuroprosthetic device may give its host new ways of manipulating physical or virtual environments through his or her body.[74] For example, it might grant enhanced control over one's existing biological body, expand one's body to incorporate new devices (such as an exoskeleton or vehicle) through body schema engineering,[75] or allow its user to control external networked physical systems such as drones or 3D printers or virtual systems or phenomena within an immersive virtual world.

Insofar as such mechanisms for motor enhancement provide proprioceptive or other sensory feedback, they would be subject to the issues noted above for neuroprostheses that provide sensory enhancement. Neuroprostheses that provide a host strengthened control over his or her body could enhance the **confidentiality** and **possession** of information by their hosts by preventing the inadvertent disclosure of information through motor actions such as speech or facial expressions. On the other hand, by extending or altering a host's body, such a device might simply create new motor avenues through which such information can be inadvertently disclosed.

c. Enhanced memory, skills, and knowledge accessible through sensory organs (exograms)

A neuroprosthetic device may give its host access to external data-storage sites whose contents can be 'played back' to the host's conscious awareness through his or her sensory organs or to real-time streams of sense data that augment or replace one's natural sense data.[76] The ability to record and play back one's own sense data could provide perfect audiovisual memory of one's experiences.[77]

Neuroprostheses that store memories, skills, and other information as exograms[78] that are external to a brain's own natural mnemonic systems face different InfoSec issues than those that store information in the form of engrams within the brain's natural mnemonic mechanisms. Information stored within engrams can be retrieved by an individual's mind without first needing

---

[74] See McGee (2008), p. 213, and Warwick (2014), p. 266.

[75] See Gladden, "Cybershells, Shapeshifting, and Neuroprosthetics: Video Games as Tools for Posthuman 'Body Schema (Re)Engineering'" (2015).

[76] See Koops & Leenes (2012), pp. 115, 120, 126.

[77] See Merkel et al., "Central Neural Prostheses" (2007); Robinett, "The consequences of fully understanding the brain" (2002); and McGee (2008), p. 217.

[78] E.g., devices of the sort described by Werkhoven, "Experience Machines: Capturing and Retrieving Personal Content" (2005), but in implantable rather than external wearable form.

to pass through sensory organs; the information appears to come 'from within' the person's own mind rather than being presented to the conscious awareness through the use of sensory organs as if the information were originating from some environment outside of the mind. The fact that the use of engrams bypasses the body's sensory systems creates different InfoSec capacities and concerns than the use of exograms that must be presented through sense organs or sensory modalities.

Because information stored as exograms can potentially take the form of conventional text, video, audio, or image files (rather than being stored as patterns of interconnection, activation functions, and learning processes within a neural network), it may be easier for unauthorized parties to access, manipulate, delete, or replace that information. On the other hand, the ability to store information in conventional digital file formats may allow the use of encryption and other security or access controls that are not possible for information stored as engrams within the brain's own neural networks – since information stored within the brain's mnemonic systems must utilize whatever form and structure the brain is designed to handle rather than whatever more 'secure' structures a security-conscious neuroprosthetic engineer might wish to impose.

Information stored in the form of exograms accessible through sensory systems would be subject to many of the same issues surrounding **integrity**, **utility**, **availability**, **confidentiality**, **authenticity**, and **possession** that currently apply, for example, to information that a user might store on a mobile device and access through earphones or a virtual reality headset. If a neuroprosthetic device can be used to forcibly present information or activate the use of skills against its user's will, then questions of **autonomy** and **rejectability** also arise.

2. Functional impairments created by a neuroprosthetic device

Below we describe some of the functional impairments that a neuroprosthetic device might create for its host at the level of his or her physical or virtual bodily interfaces with the environment and the impact that these impairments might have on the information security of the host-device system.[79]

a. Loss of control over sensory organs

A neuroprosthesis may deny its host or user direct control over his or her sense organs.[80] Technologically mediated sensory systems may be subject to

---

[79] See Gladden, "Neural Implants as Gateways to Digital-Physical Ecosystems and Posthuman Socioeconomic Interaction" (2016).

[80] Koops & Leenes (2012), p. 130.

noise, malfunctions, and manipulation or forced sensory deprivation or over-load occurring at the hands of 'sense hackers.'[81]

A neuroprosthetic device that intentionally deprives its host of control over his or her sense organs raises questions of the **rejectability** of sense data and may impair the host's exercise of agency and thus his or her **autonomy**. The **availability, integrity, utility,** and **authenticity** of information provided by the host's augmented sense organs will depend on the technical capacities and motives of whatever external agents control the design or operation of such a neuroprosthetic device.

b. Loss of control over motor organs

A neuroprosthetic device may impede its host or user's control over his or her motor organs.[82] A host's body may no longer be capable, for example, of speech or movement, or the control over his or her speech or movements may be assumed by some external agency.

A neuroprosthetic device that intentionally deprives its host of control over his or her motor organs may prevent that person from inadvertently (or even purposefully) disclosing information through the use of speech, facial expressions, typing, or other physical means, thereby enhancing the **confiden-tiality** and **possession** of information. Meanwhile, the use of such a device may impair the information security of outside parties who, for example, interact with the host in conversation, listen to the host giving a lecture, or read a message that was typed and sent by the host: such individuals might assume that the information was conveyed to them intentionally by the host, when in fact it may have been conveyed against the host's will by some external agent who was controlling the host's motor activity through the neuropros-thetic device. This would result in a loss of **authenticity** of the information shared by the 'host,' from the perspective of those who received it.

c. Loss of control over other bodily systems

A neuroprosthetic device may impact the functioning of internal bodily processes such as respiration, cardiac activity, digestion, hormonal activity, and other processes that are already affected by existing implantable medical devices.[83]

Insofar as a neuroprosthetic device interfaces directly with such biological systems and processes, it may gather, store, utilize, and transmit data about

---

[81] In Hansen & Hansen, "A Taxonomy of Vulnerabilities in Implantable Medical Devices" (2010), there is introduced and discussed the hypothetical case of a poorly designed prosthetic eye whose internal computer can be disabled if the eye is presented with a particular pattern of flashing lights.

[82] Gasson (2012), pp. 14-16.

[83] See McGee (2008), p. 209, and Gasson (2012), pp. 12-16.

them that must be secured in order to avoid a loss of **confidentiality** and **possession** of the information. By affecting the body's basic biological processes, a device may impact the brain's ability to receive, generate, store, transmit, and consciously experience information and may thus indirectly affect the **availability** and **utility** of information available to its host's mind.

d. Other biological side-effects

A neuroprosthetic device may be constructed from components that are toxic or deteriorate in the body,[84] may be rejected by its host, or may be subject to mechanical, electronic, or software failures that harm their host's organism.

Depending on the nature and severity of such effects, negative impacts could result for a host's **autonomy** and the **rejectability, integrity, utility, availability, authenticity**, and **distinguishability** of information experienced by a host or user. On the other hand, if an advanced neuroprosthetic device is only able to function for a limited period of time before its connection to the neural circuitry of its host breaks down and the device ceases to function, this behavior could function as a sort of safeguard that limits the long-term possibilities for the device to contribute to a loss of the **confidentiality** or **possession** of information.

E. The host as social and economic actor

1. Functional capacities created by a neuroprosthetic device

Below we describe some of the new functional capacities that a neuroprosthetic device might provide that allow its host or user to connect to, participate in, contribute to, and be influenced by social relationships and structures and networks of economic exchange. We also note the potential impact that these capacities might have on the information security of the host-device system.[85]

a. Ability to participate in new kinds of social relations

A neuroprosthetic device may grant its host or user the ability to participate in new kinds of technologically mediated social relations and structures that were previously impossible, perhaps including new forms of merged agency[86] or cybernetic networks that display utopian (or dystopian) characteristics that are not possible for non-neuroprosthetically-enabled societies.[87]

The creation of novel kinds of social relationships may create new avenues for a host or user to inadvertently disclose information, thereby damaging its

---

[84] McGee (2008), pp. 213-16.

[85] See Gladden, "Neural Implants as Gateways to Digital-Physical Ecosystems and Posthuman Socioeconomic Interaction" (2016).

[86] See McGee (2008), p. 216, and Koops & Leenes (2012), pp. 125, 132.

[87] See Gladden, "Utopias and Dystopias as Cybernetic Information Systems" (2015).

confidentiality and his or her **possession** of it. It may also provide new means for external agents to disrupt or influence the host or user's acquisition and use of information, not through manipulation of a device's components or systems but by using the device in its intended fashion to interact socially with its host or user and undermining his or her information security through the nature and contents of those social interactions (which might involve social engineering[88]). This could indirectly impact the **availability** and **utility** of information for the user and may also potentially undermine his or her agency and thus **autonomy**.

b. Ability to share collective knowledge, skills, and wisdom

Neuroprostheses may link hosts or users in a way that forms communication and information systems[89] that can generate greater collective knowledge, skills, and wisdom than are possessed by any individual member of the system.[90]

On the one hand, using a neuroprosthetic device to store information in communal systems that make their contents freely accessible to other human minds clearly eliminates a host's ability to maintain the **confidentiality** and **possession** of that information. On the other hand, by drawing information from 'open-source' repositories whose maintenance and editing are crowdsourced to myriad minds that are continuously identifying and rectifying errors and which provide checks and balances to counteract one another's biases, it may be possible for such 'neuroprosthetically enabled wikis' to maintain a self-healing and self-correcting state that offers greater **availability**, **integrity**, and **utility** of information than that possible from a static source developed by a single author.

c. Enhanced job flexibility and instant retraining

By facilitating the creation, alteration, and deletion of information stored in engrams or exograms, a neuroprosthetic device may allow its host to download new knowledge or skills or instantly establish relationships for use in a new job.[91]

---

[88] See Rao & Nayak (2014), pp. 307-23; Sasse et al., "Transforming the 'weakest link'—a human/computer interaction approach to usable and effective security" (2001); and Thonnard, "Industrial Espionage and Targeted Attacks: Understanding the Characteristics of an Escalating Threat" (2012).

[89] See McGee (2008), p. 214; Koops & Leenes (2012), pp. 128-29; Gasson (2012), p. 24; and Gladden, "Enterprise Architecture for Neurocybernetically Augmented Organizational Systems" (2016).

[90] See Wiener, *Cybernetics: Or Control and Communication in the Animal and the Machine* (1961), loc. 3070ff., 3149ff., and Gladden, "Utopias and Dystopias as Cybernetic Information Systems" (2015).

[91] See Koops & Leenes (2012), p. 126, and Gladden, "Neural Implants as Gateways to Digital-Physical Ecosystems and Posthuman Socioeconomic Interaction" (2016).

A neuroprosthetic device that allows its host to enhance his or her socio-economic position by continuously upgrading his or her skills, enhancing his or her job performance, and moving into ever more desirable and rewarding professions and positions may provide the host with resources (including financial, informational, and human resources and access to new technologies embodied in hardware, software, and services) that allow him or her to enhance and strengthen his or her information security, including his or her **autonomy**, the **confidentiality** and **possession** of information already in his or her control, the **availability** of new kinds of information, and enhanced tools for extracting **utility** from information.

d. Enhanced ability to manage complex technological systems

By providing a direct interface to external computers and mediating its user's interaction with them,[92] a neuroprosthetic device may grant an enhanced ability to manage complex technological systems that can be employed, for example, in the production or provisioning of goods or services or the management of digital ecosystems and environments that utilize ubiquitous computing and are integrated into the Internet of Things.[93]

By giving the user of a neuroprosthesis enhanced capacities for acquiring and managing information and controlling his or her environment, a device may offer the user increased **availability**, **utility**, **confidentiality**, **possession**, and **rejectability** of information.

e. Enhanced personal and professional decision-making

By analyzing data, offering recommendations, and alerting its user to potential cognitive biases, a neuroprosthetic device may enhance the user's ability to execute rapid and effective personal and professional decision-making and transactions.[94]

By enhancing its user's ability to avoid the effects of internal biases and to identify and counteract intentional or inadvertent efforts by others to manipulate the user through social interaction, such a neuroprosthesis may enhance its user's agency and **autonomy** and help prevent him or her from inadvertently making decisions or undertaking actions that would undermine the **confidentiality** or **possession** of the user's information. It may also lead the user to make decisions that will eventually put him or her in a position to enjoy greater **availability** and **utility** of information.

---

[92] McGee (2008), p. 210.

[93] See McGee (2008), pp. 214-15, and Gladden, "Enterprise Architecture for Neurocybernetically Augmented Organizational Systems" (2016).

[94] See Koops & Leenes (2012), p. 119.

f. Store of monetary value

By storing cryptocurrency keys within its internal memory, an implanted neuroprosthesis may allow its host to house digital money directly within his or her brain that he or she can spend on demand.[95]

The use of a neuroprosthetic device to store information that has direct monetary value – rather than simply confidential personal or professional information that an unauthorized party might steal and attempt to convert into money through its sale or through blackmail of the host – creates an enticing new target for criminals and a new kind of information that must be carefully secured. For many users, the **possession** and **confidentiality** of such financial information would take priority and must be safeguarded, even if it means reducing the **availability** and **utility** of the information to the user.

A neuroprosthetic device that can be used directly to purchase goods and services and engage in other forms of economic exchange may give its user new tools for acquiring, utilizing, and securing information, thereby increasing the **availability** and **utility** of information as well as its **confidentiality** and **possession**.

g. Qualifications for specific professions and roles

Neuroprostheses may provide persons with abilities that enhance job performance in particular fields[96] such as computer programming, art, architecture, music, economics, medicine, information science, e-sports, information security, law enforcement, and the military. This may initially provide a competitive advantage to individuals using certain kinds of neuroprosthetic devices while not excluding from such work those who lack neuroprostheses. However, it is expected that as the use of elective neuroprostheses becomes more commonplace and employers' expectations for employees' neural integration into digital workplace systems grow, in some professions possession of neuroprostheses may become a basic requirement for employment that excludes from consideration potential workers who do not possess such devices.[97]

Insofar as individuals' use of advanced neuroprosthetic devices is a necessary and important aspect of their professional work, it can be expected that such employees' workplaces and employers will create and maintain robust institutional support systems for the users of such devices, which may include the attention of InfoSec professionals dedicated to securing the information contained in these host-device systems. Such support structures may provide

---

[95] See Gladden, "Cryptocurrency with a Conscience: Using Artificial Intelligence to Develop Money that Advances Human Ethical Values" (2015).

[96] Koops & Leenes (2012), pp. 131-32.

[97] McGee (2008), pp. 211, 214-15, and Warwick (2014), p. 269.

employees with stronger mechanisms for ensuring the **availability**, **utility**, **integrity**, **confidentiality**, and **possession** of information than they could obtain on their own if they acquired and utilized their neuroprosthetic devices solely as ordinary consumers and non-institutional users of such devices. On the other hand, by allowing their employers to exercise at least some of the responsibility for maintaining, managing, and securing their neuroprosthetic devices, such users might instead find that an employer claims and acquires access to a user's personal information produced or accessible through the neuroprosthesis, thereby reducing the **confidentiality** and **possession** of information by the user and potentially raising questions for the user about the extent to which the information's **availability**, **integrity**, and **authenticity** can be relied upon.

2. Functional impairments created by a neuroprosthetic device

Below we describe some of the functional impairments that a neuroprosthetic device might create for its host or user at the level of his or her social and economic relationships and activity and the impact that these impairments might have on the information security of the host-device system.[98]

a. Loss of ownership of one's body and intellectual property

A neuroprosthetic device that is being leased by its human host rather than having been purchased would not belong to the host. Moreover, even a neuroprosthesis that has been purchased by its host might, under some legal regimes, potentially be subject to seizure by an outside party in some circumstances (e.g., after a declaration of bankruptcy by the host). Depending on the leasing or licensing terms, intellectual property produced by a neuroprosthetic device's host or user (including thoughts, memories, or speech) may be partly or wholly owned by the device's manufacturer or provider.[99]

This may result in binding limits on the **confidentiality**, **possession**, **availability**, and **utility** of information that can be enforced by the device's manufacturer or provider through either legal or technical means. The manufacturer or provider may also have the legal right and technical ability to forcibly present to the user's conscious or subconscious awareness explicit advertisements, product placements edited into the user's sense data, or other commercial information that undermines the **rejectability** and perhaps **distinguishability** and **authenticity** of information received through the device. The fine print of the leasing, licensing, or even purchase agreement may also specify that the device's manufacturer or provider has the legal right to utilize the

---

[98] See Gladden, "Neural Implants as Gateways to Digital-Physical Ecosystems and Posthuman Socioeconomic Interaction" (2016).

[99] See Gladden (2017), pp. 248-49.

device to gather on an ongoing basis information about its host or user (including information about his or her mental and biological processes), which the company can either use internally for its own purposes or perhaps rent or sell to other companies for their own ends. This would have the effect of significantly reducing the **confidentiality** and **possession** of personal information by the host or user. On the other hand, by maintaining an ongoing financial relationship with the device's manufacturer or provider, the user may be able to make use of physical maintenance services, software updates and upgrades (including regular updating of anti-malware and other security software), and other services provided by that firm which enhances the **confidentiality, possession, availability, utility,** and **integrity** of information experienced through the device.

b. Creation of financial, technological, or social dependencies

The host or user of a neuroprosthesis may no longer be able to function effectively without the device[100] and may become dependent on its manufacturer for hardware maintenance, software updates, and data security and on specialized medical care providers for diagnostics and treatment relating to the device.[101] A user may also require regular device upgrades in order to remain competitive in certain jobs for which the possession and expert use of such a device is a job requirement. High switching costs may make it impractical for a host to shift to a competing producer's device after he or she has installed an implant and committed to its manufacturer's particular digital ecosystem.

If the host or user of a neuroprosthetic device is likely to suffer psychological, biological, financial, professional, or social damage without such ongoing specialized support from a company, this creates a power relation in which the host or user is in a position of dependency (or even subjugation) and in which he or she may be willing to accept an exploitative situation in which the **confidentiality** and **possession** of his or her information is compromised by the company, his or her autonomy is diminished, and the **availability,**

---

[100] Koops & Leenes (2012), p. 125.

[101] See McGee (2008), p. 213. Brain scarring is a significant problem with neuroprostheses that involve electrodes implanted in the brain, and the administration before, during, and after implantation surgery of immunosuppressive drugs that reduce the wound-healing response has been found to reduce scarring and cortical edemas; see Polikov et al., "Response of brain tissue to chronically implanted neural electrodes" (2005), for a discussion of such issues. The possibility that the host of an implanted advanced neuroprosthetic device might become dependent throughout the rest of his or her life on the device's manufacturer (or another commercial entity) for a regular supply of potentially expensive and proprietary immunosuppressive drugs or other specialized medications is a theme that has been explored, e.g., by futurologists and the creators of science fiction works; for an analysis of one fictional depiction, see Maj, "Rational Technotopia vs. Corporational Dystopia in 'Deus Ex: Human Revolution' Gameworld" (2015).

utility, **integrity**, and **rejectability** of information is subject to the whims (likely driven by financial considerations) of the company.

c. Subjugation of the host to manipulation by external agency

Instead of merely impeding its host or user's ability to possess and exercise agency, a neuroprosthesis may subject its host to control by some external agency. This could occur, for example, if the host's memories, emotions, or volitions were manipulated by means of the device[102] or if the host joined with other minds to create a new form of social entity that possessed some shared agency.[103]

Such a situation would impair the host's **autonomy** and could be exploited to undermine the **confidentiality** and **possession** of the host's information. Depending on the level of access to the host's information that is gained by the external agent, the **authenticity**, **integrity**, **availability**, and **utility** of the host's information could also be imperiled.

d. Social exclusion and fragmentation and employment discrimination

The use of kinds of neuroprostheses that are considered by a particular society to be of a 'suspicious' or 'undesirable' nature and whose presence and operation is detectable to parties other than their host or user may potentially result in the shunning or mistreatment of such hosts or users[104] by those who question or actively oppose the use of such devices.[105] Hosts or users of such neuroprosthetic devices may find themselves formally or informally excluded from certain kinds of organizations and social relationships, or they may simply avoid certain kinds of relationships and situations in order to spare themselves the embarrassment or discomfort that might result from such interactions. Possession of some kinds of neuroprostheses may exclude their hosts from employment in roles where 'natural,' unmodified workers are considered desirable or even required (e.g., for liability or security reasons).

It is also expected that some kinds of advanced neuroprostheses will so radically transform their users' mechanisms for communicating and interact-

---

[102] Gasson (2012), pp. 15-16.

[103] See McGee (2008), p. 216; McIntosh (2010); Roden (2014), p. 39; and Gladden, "Utopias and Dystopias as Cybernetic Information Systems" (2015).

[104] Koops & Leenes (2012), pp. 124-25.

[105] For example, anecdotal accounts have already been reported of physical harassment and exclusion from places of business of individuals wearing external sensory prostheses designed to generate visual augmented reality. See Greenberg, "Cyborg Discrimination? Scientist Says McDonald's Staff Tried To Pull Off His Google-Glass-Like Eyepiece, Then Threw Him Out" (2012), and Dvorsky, "What may be the world's first cybernetic hate crime unfolds in French McDonald's" (2012).

ing socially that they will eventually lose the desire and even ability to communicate with human beings who do not possess the relevant sort of neuroprostheses; in this way, humanity as it exists today may fragment into numerous mutually incomprehensible 'posthumanities' that share a geographical home on this planet but whose societies and civilizations occupy disjoint psychological, cultural, and technological spaces that do not intersect or overlap.[106] Such a splintering and narrowing of societies may possibly weaken the solidarity with other human beings felt by users of some kinds of advanced neuroprostheses.[107]

If the users of certain kinds of neuroprostheses were, in essence, to withdraw from 'normal' human society and develop new societies accessible only to those who share similar technological augmentation, interests, and philosophies, one side-effect of that growing distance and insulation from unaugmented human society could be an increase in the **confidentiality** and **possession** of information by the users of such technologies, insofar as the ability of unaugmented humans to initiate some kinds of attacks (such as social engineering efforts) against them might be significantly curtailed. At the same time, the users of such neuroprosthetic devices might find that their voluntary or involuntary distancing from the rest of humanity separates them legally, politically, commercially, socially, or technologically from InfoSec systems and mechanisms that are available to other human beings, thereby potentially putting at risk the **integrity**, **availability**, **confidentiality**, and **possession** of the users' information.

e. Vulnerability to data theft, blackmail, and extortion

A hacker, piece of malware, or other agent may be able to steal data contained in a neuroprosthetic device or use the device to gather data (potentially including the contents of thoughts, memories, or sensory experiences)[108] that could be used for blackmail, extortion, corporate espionage, or terrorism targeted against the device's host or user or other individuals or institutions. Such an attacker could either carry out a one-time theft of information or embed software (or even hardware) in the device that allows ongoing access and the ability to utilize the device's features and components as an instrument for information-gathering and surveillance, regardless of whether they were designed to be employable for such purposes.

The minds, personalities, interests, motivations, and values of all human beings differ to a large extent, which means that the authors of certain kinds

---

[106] See McGee (2008), pp. 214-16; Warwick (2014), p. 271; and Rubin, "What Is the Good of Transhumanism?" (2008).

[107] Koops & Leenes (2012), p. 127.

[108] See McGee (2008), p. 217; Koops & Leenes (2012), pp. 117, 130; and Gasson (2012), p. 21.

of social engineering attacks on high-value targets must take the time to learn about the subject of their intended operation and develop a customized plan of attack – and it cannot be known for certain in advance of the attack whether or not it will succeed and what unexpected obstacles might arise during its attempted execution.[109] If the target of such an attack possesses an advanced neuroprosthetic device, this may give the attacker a means of planning and executing the attack that depends solely on technical and technological factors (which it may be possible to analyze carefully in advance) rather than social and psychological ones. Exploitable vulnerabilities in a particular model of neuroprosthetic device that have been identified by would-be attackers may place at risk all human beings who possess that particular model of device, regardless of the otherwise great psychological, cultural, and professional dissimilarities between them.

Depending on the exact purpose and nature of such an attack, it may have the potential to undermine the **confidentiality** and **possession** both of the host or user's information and the information of other parties that can be compromised by means of the neuroprosthetic device (e.g., by using a neuroprosthesis implanted in one person to eavesdrop on a separate individual who happens to be nearby). It may also compromise the **authenticity, integrity, availability, distinguishability**, and **utility** of information and be employed to undermine the host or user's **autonomy**.

V. Conclusion

In this chapter we have explored a two-dimensional conceptual framework for cognitional security that comprises nine essential information security goals for neuroprosthetic devices and host-device systems (confidentiality, integrity, availability, possession, authenticity, utility, distinguishability, rejectability, and autonomy) and examines potential impacts on the pursuit of those goals as observed at three different levels (which consider a device's host understood as sapient metavolitional agent, embodied embedded organism, and social and economic actor). In the following chapters we will draw on this cognitional security framework to consider important practical issues relating to the development and implementation of InfoSec plans for advanced neuroprostheses – namely, the formulation of particular information security roles and responsibilities and the design and use of preventive, detective, and corrective controls.

---

[109] See Rao & Nayak (2014), pp. 307-23, and Sasse et al. (2001).

Part II

Security Practices and Mechanisms for Advanced Neuroprosthetics

Chapter Five

InfoSec Roles and Responsibilities for the Securing of Neuroprosthetic Systems

Abstract. This chapter describes how responsibilities for planning and implementing information security practices and mechanisms are typically allocated among individuals filling particular roles within an organization. It then investigates the unique forms that these InfoSec roles and responsibilities can take when the focus of their activities is ensuring information security for advanced neuroprostheses and their human hosts.

I. Introduction

The process of ensuring information security for a neuroprosthetic system involves the combined effort of many individuals carrying out different activities at a range of levels, regardless of whether the system is being used within a large institutional setting or by a single consumer. Here we consider classic descriptions of key information security roles and responsibilities and explore the ways in which these roles may need to be adjusted or complemented by newly defined roles and responsibilities that are uniquely important for an advanced neuroprosthetic system.

II. Overview of security activities by SDLC phase

Information security roles and responsibilities relate to the execution of particular activities; it is thus useful to describe those activities before considering the individuals who will carry them out. The many activities needed to ensure information security can be organized according to their place within the system development life cycle (SDLC), which spans the period from when the first idea is raised about acquiring or developing a new system through the time when that system, no longer 'new,' is eventually removed from service.[1]

---

[1] *NIST Special Publication 800-100: Information Security Handbook: A Guide for Managers* (2006), pp. 19-25.

In principle, a system development life cycle can be quite brief; it may only last days, weeks, or months, if the new system is small in scope and complexity and is being adopted only for a limited time to fulfill some *ad hoc* purpose. Although the key elements of the SDLC should still be present, in such a case they might appear in only a brief and simplified form. In the case of advanced neuroprostheses, the costs, risks, and complexity involved with implanting such devices into and removing them from human hosts means that their SDLC – and in particular, the operations and maintenance phase of the SDLC during which the device is functioning within its human host – may be more likely to span years or even decades.

Below we consider an SDLC comprising five main phases – the initiation, development and acquisition, implementation, operations and maintenance, and disposal phases – as they relate to advanced neuroprostheses. Here we draw extensively on the description of these phases found in Chapter Three of *NIST Special Publication 800-100: Information Security Handbook: A Guide for Managers*, produced by the National Institute of Standards & Technology. We have integrated into that general framework a number of additional activities (such as awareness and training and the interconnection of systems) that are discussed elsewhere in *NIST SP800-100* but not explicitly as components of the SDLC.[2]

A. Initiation phase

1. Determination of needs

During the initiation phase, the needs that should be filled by the neuroprosthetic device are specified.[3] For example, the device may be intended to function therapeutically in treating a particular medical condition or it may be designed to augment the abilities of a human user by providing some psychological or physical enhancement.[4] The required capacities and features that the device should possess are described at a general level.[5]

---

[2] In addition to the general description of an SDLC in the context of information security that is found in *NIST SP 800-100* (2006), see Gladden, "Managing the Ethical Dimensions of Brain-Computer Interfaces in eHealth: An SDLC-based Approach" (2016), for a description of an SDLC with five phases (analysis and planning; design, development, and acquisition; integration and activation; operation and maintenance; and disposal) applicable to brain-computer interface technologies utilized in e-health.

[3] *NIST SP 800-100* (2006), p. 22.

[4] See Chapter One of this book for a discussion of different potential purposes for advanced neuroprosthetic devices and systems.

[5] *NIST SP 800-100* (2006), p. 22.

2. Security categorization

Next the kinds of information that the device will receive, generate, store, and transmit are identified. The relevant types of information can then be categorized using schemes such as those found in *NIST SP 800-60* and *FIPS 199*, thereby describing the nature and sensitivity of the information that the device will possess.[6] Knowing the kinds of information that the device will handle and the sensitivity of that information is essential in order to make effective decisions regarding the information security policies, practices, and mechanisms that will be employed. For example, a neuroprosthetic device that can access and record the contents of its host's mental processes or which can be used as a surveillance device to record nearby conversations will require a much different InfoSec plan than a neuroprosthesis whose sole function is to stimulate the release of hormones within its host's body according to some regular schedule.

3. Preliminary risk assessment

A preliminary assessment is undertaken to "define the threat environment in which the system or product will operate."[7] A neuroprosthesis that will be used within a controlled clinical environment for medical research purposes and which cannot be modified by its host presents different risks than a mass-produced consumer neuroprosthetic device that will be used by individuals in their homes and workplaces and which can be modified by its hosts. A neuroprosthetic device for use by military personnel in battlefield operations – or cyberwarfare – will encounter still different risks.[8]

In the case of some kinds of neuroprostheses, it may be difficult or impossible to specify a 'typical' or 'normal' operating environment, if that environment is influenced or determined by the cognitive processes of a host or user and the contents or his or her mental activity; in such a situation, the operating environments presented by the minds of two different hosts could differ as greatly as do the memories, experiences, dreams, desires, and fears of two different people, and the operating environment presented by the mind of even a single host could change radically depending on the host's current emotional state and whether, for example, he or she is asleep or awake.

---

[6] *NIST SP 800-100* (2006), p. 22, and *FIPS PUB 199: Standards for Security Categorization of Federal Information and Information Systems* (2004). See Chapter One of Gladden, *Neuroprosthetic Supersystems Architecture* (2017), for a device ontology that can be used to identify and classify such kinds of information.

[7] *NIST SP 800-100* (2006), p. 22.

[8] On potential military use, see Schermer, "The Mind and the Machine. On the Conceptual and Moral Implications of Brain-Machine Interaction" (2009); Brunner & Schalk, "Brain-Computer Interaction" (2009); and Chapter Four of Gladden (2017).

B. Development and acquisition phase

1. Requirement analysis and development

Based on the decisions and information produced during the initiation phase, a more detailed analysis of requirements for a device (including information security requirements) is now carried out and specifications for the device are developed.[9]

2. Risk assessment

A more in-depth risk assessment is conducted that is based on the device's specifications and intended (or possible) operating environments and informed by functional, legal, and ethical considerations.[10] This process also includes the identification of vulnerabilities[11] and threats[12] and the development of recommendations for security controls.[13]

3. Cost considerations and reporting

In many ways it would be easier to develop effective InfoSec plans if cost were not an issue and unlimited financial and human resources could be dedicated to information security. In reality, the resources available to support information security are always limited and sometimes quite small relative to the scope, complexity, and importance of a project. InfoSec decision-makers often find themselves developing not the "best information security plan possible" but rather the "best information security plan possible given the resources available."

Cost considerations must thus be carefully and realistically considered as part of the development and acquisition phase.[14] This is especially true for a neuroprosthesis that may be permanently implanted in a human being, as analysis might indicate that the cost of ensuring information security for the

---

[9] *NIST SP 800-100* (2006), p. 22. See Chapter Two of this text for a discussion of vulnerabilities of neuroprosthetic devices. For vulnerabilities of IMDs generally, see Hansen & Hansen, "A Taxonomy of Vulnerabilities in Implantable Medical Devices" (2010).

[10] *NIST SP 800-100* (2006), p. 22. For an overview of ethical issues with ICT implants – many of which are relevant for advanced neuroprosthetics – see Hildebrandt & Anrig, "Ethical Implications of ICT Implants" (2012). For ethical issues in information security more generally, see Brey, "Ethical Aspects of Information Security and Privacy" (2007).

[11] *NIST SP 800-100* (2006), p. 88.

[12] *NIST SP 800-100* (2006), p. 87. For a discussion of threats to neuroprosthetic devices, see Chapter Two of this text and Denning et al., "Neurosecurity: Security and Privacy for Neural Devices" (2009). For threats to IMDs generally, see Halperin et al., "Security and privacy for implantable medical devices" (2008); for threats to BSNs, see Cho & Lee, "Biometric Based Secure Communications without Pre-Deployed Key for Biosensor Implanted in Body Sensor Networks" (2012).

[13] *NIST SP 800-100* (2006), p. 90.

[14] *NIST SP 800-100* (2006), p. 90.

device may be expected to increase or decrease significantly over the period of years or decades in which the device may be in use. Difficult and complex situations may arise, for example, if a neuroprosthetic device is implanted in a human being by the host's employer and the information security costs are borne by the institution as long as the person is an employee, but financial and technical support ceases upon termination of the host's employment.

4. Security planning

A comprehensive device or system security plan (including documentation for all security controls[15]) is developed to guide the subsequent phases and processes. Complementary resources such as training materials and manuals for administrators and users are also prepared.[16]

The system security plan does not simply address the InfoSec performance of the implantable neuroprosthetic unit itself nor only when operating under nominal conditions. The plan should also incorporate information technology contingency planning for dealing with anomalous situations[17] and should consider the InfoSec performance of the joint systems that the neuroprosthetic unit will create through its integration and interconnection with other systems (including the outcomes that will occur, for example, if an emergency disconnection of the neuroprosthetic unit from its complementary systems must be performed[18]).

5. Security control development

The security controls described in the system security plan are now designed and implemented.[19] Note that insofar as possible, all security controls are implemented within the system before the system itself is implemented within the production environment in which it will operate; while it may be possible to implement all of the logical (and perhaps physical) controls prior to the system's deployment, it may only be possible to implement the full

---

[15] For the importance of documentation especially for enterprise information systems, see Chaudhry et al., "Enterprise Information Systems Security: A Conceptual Framework" (2012). Some advanced neuroprosthetic devices might indeed be used in such a context as components of enterprise information systems; see Gladden (2017).

[16] *NIST SP 800-100* (2006), pp. 23, 26-34.

[17] *NIST SP 800-100* (2006), pp. 78-82. See Chapter Three of this text for a discussion of the need to access neuroprosthetic devices during health emergencies experienced by their human host. See also Clark & Fu, "Recent Results in Computer Security for Medical Devices" (2012); Rotter & Gasson, "Implantable Medical Devices: Privacy and Security Concerns" (2012); and Halperin et al. (2008) – all of whom raise the issue of emergency access to IMDs. Halperin et al., especially, consider this question in detail.

[18] *NIST SP 800-100* (2006), p. 52.

[19] *NIST SP 800-100* (2006), pp. 23, 113-23.

range of administrative controls after the system has been deployed and all of the relevant personnel and offices that will manage the use of the administrative controls are fully engaged with the use of the system.[20] Particular kinds of security controls that may be relevant for advanced neuroprosthetic devices will be described in more detail in later sections of this book.

6. Developmental security test and evaluation

Insofar as possible, all security controls are tested to determine their effectiveness prior to the system's deployment so that weaknesses and problems can be identified, changes and improvements can be made, and the system can then be retested. Specific scenarios (such as electronic hacking and social engineering attempts, the presence of malware, operator error, and the failure of equipment or power supplies) may be prepared and their effects played out in order to evaluate the system's response.[21]

For some kinds of neuroprosthetic devices, it will not be possible to robustly test a device's InfoSec features and performance until after the device has been implanted in its human host, insofar as the cognitive and physical connections between the device and host are among the key processes whose InfoSec performance must be tested and the integration of the device into the host's psychological processes and biological systems cannot easily be simulated prior to the device's implantation.

7. Other planning components

Other planning components that are carried out during the development and acquisition phase include considering the kinds of contracts[22] and agreements that will need to be developed between suppliers of raw materials, component and device manufacturers, software designers, distributors, organizations providing implantation surgery and medical support, and end users.

C. Implementation phase

1. Security test and evaluation

Final pre-deployment testing of a system's InfoSec performance is undertaken in part to gather data about the system's baseline functioning and in part to ensure that the final set of security policies, practices, and mecha-

---

[20] *NIST SP 800-100* (2006), p. 23.

[21] *NIST SP 800-100* (2006), p. 23.

[22] *NIST SP 800-100* (2006), p. 23.

nisms utilized by the system meets all relevant legal and regulatory requirements and satisfies ethical standards.[23] The final pre-deployment testing of
the system's InfoSec performance should not test the system in isolation but,
insofar as possible, should test the interaction of the system with all of the
other systems with which it will be interconnected or on which it may be
dependent.[24]

In the case of an advanced neuroprosthetic device, legal, regulatory, and
ethical issues surrounding the privacy of personal health information will be
a particular concern. Even if the device is being used by its host as an entertainment device or to enhance work productivity – and is not thought of primarily as a 'medical device' – the system may still be interacting with the
host's body and gathering data in such a way that subjects it to potentially
restrictive and intense legal and regulatory regimes.

2. Inspection and acceptance

The operator who is accepting the neuroprosthetic device for use verifies
that all of the InfoSec features and capacities that were required as part of the
device's specifications are indeed present and functional in the finished product as supplied by the developer or manufacturer.[25]

3. System integration and installation

The device is deployed as the necessary steps are taken to "Integrate the
system at the operational site where it is to be deployed for operation" and to
"Enable security control settings and switches in accordance with vendor instructions and proper security implementation guidance."[26]

In the case of many advanced neuroprostheses, 'system integration and
installation' constitutes the process of implanting a device into its human
host, integrating the device with the host's neural circuitry, and activating the
device.[27]

---

[23] *NIST SP 800-100* (2006), p. 23. On the role of regulators and regulation in the development and
use of neuroprostheses, see McCullagh et al., "Ethical Challenges Associated with the Development and Deployment of Brain Computer Interface Technology" (2013); Patil & Turner, "The Development of Brain-Machine Interface Neuroprosthetic Devices" (2008); Kosta & Bowman, "Implanting Implications: Data Protection Challenges Arising from the Use of Human ICT Implants"
(2012); and Gladden, "Information Security Concerns as a Catalyst for the Development of Implantable Cognitive Neuroprostheses" (2016).

[24] *NIST SP 800-100* (2006), p. 23.

[25] *NIST SP 800-100* (2006), p. 23.

[26] *NIST SP 800-100* (2006), p. 23.

[27] See Chapter One of this text for a general discussion of the integration of neuroprosthetic
devices into the neural circuitry of their human host and Chapter One of Gladden (2017) for a
device ontology that can be used to specify such interconnections for a particular device in more

4. Security certification

Government agencies or large corporations may possess a 'security certification' granted either by a senior official within the organization or by some external body certifying that the organization follows best practices for information security and that its systems utilize security controls, safeguards, and countermeasures that are sufficient to keep the organization's information secure.[28]

Existing certifications of this sort may need to be updated to include advanced neuroprosthetic systems newly adopted by an organization. For example, a government agency that has begun to employ personnel who possess artificial eyes that are capable of recording and transmitting live video would need to ensure that the addition of such devices to the workplace ecosystem does not compromise organizational information security. In the case of neuroprosthetic devices designed as consumer electronics products for use by individuals, the manufacturer may wish to reassure (potential) customers by obtaining certification from an independent body that – when used in accordance with the manufacturer's instructions – the device will operate in a secure manner and will not compromise a user's information security.

5. Security accreditation

Government agencies or large corporations whose personnel utilize neuroprosthetic devices may maintain a process of security accreditation by which a new kind of neuroprosthetic device is approved for use in receiving, generating, storing, or transmitting specific kinds of information.[29] Until a particular kind of neuroprosthetic device has received such accreditation, its hosts or users may be barred from accessing designated kinds of information or filling certain types of roles within the organization.[30]

Given the fact that the secure and effective functioning of an advanced neuroprosthesis depends on the unique nature and extent of its integration with its human host and that the same type of neuroprosthetic device may operate very differently when implanted in different human hosts, some organizations may require a process of security accreditation not for each general *type* of neuroprosthetic device but for each specific host-device pairing: a device may create a secure host-device system as implanted and functioning in one human host, while in a different host the same type of device would not create such a system.

---

detail.

[28] *NIST SP 800-100* (2006), p. 23.

[29] *NIST SP 800-100* (2006), p. 24.

[30] See Chapter Four of this text for a discussion of potential employment discrimination relating to advanced neuroprosthetic devices.

D. Operations and maintenance phase

1. Configuration management and control

A neuroprosthetic device may be capable of operating in many different configurations. Variation between configurations may arise from the installation of different software; the selection of different settings within software; the addition, removal, or adjustment of hardware components; or the adjustment of the device's relationship to and interconnection with its human host or external systems. In order to maximize information security, it is essential to choose the correct initial configuration for the device, change the configuration as needed, respond to environmental conditions or operational needs, and keep an accurate record of all changes to the configuration that have been made.[31]

For some kinds of neuroprosthetic devices, it may be difficult or impossible to clearly define different 'configurations' that have been employed by a device. While many devices will include hardware or software settings that can be discretely adjusted and whose current configuration can be simply and precisely recorded, other neuroprostheses might utilize, for example, electronic physical neural networks – or even living biological components – whose states are continually changing and evolving and are impossible to fully and precisely define.[32]

2. Continuous monitoring

After deployment of a neuroprosthetic device, continuous monitoring is required in order to determine whether its actual operating conditions and environment are consistent with the conditions for which it was designed and to ensure that relevant security policies, practices, and mechanisms are functioning as intended.[33] Monitoring may include the automated generation and analysis of logfiles as well as manual audits and assessments that may be carried out with the use of particular software.[34]

The detection of anomalous conditions (whether a hardware failure or intentional attack on the system) may trigger an incident response including the deployment of countermeasures designed for "containment, eradication, and recovery" from the problem.[35] In the case of advanced neuroprostheses, the kinds of countermeasures that can be employed will be constrained by

---

[31] *NIST SP 800-100* (2006), p. 24.
[32] See the device ontology presented in Chapter One of Gladden (2017) for a discussion of such physical neural networks.
[33] *NIST SP 800-100* (2006), p. 24.
[34] *NIST SP 800-100* (2006), p. 24.
[35] *NIST SP 800-100* (2006), pp. 124-29.

legal, regulatory, and ethical considerations that place an obligation on operators to immediately and effectively address any incident that affects a device implanted within a human being – while simultaneously requiring that the health, safety, and well-being of the device's human host not be compromised by any InfoSec countermeasures utilized to respond to the incident affecting the device.[36]

E. Disposal phase

1. Information preservation

The long-term disposition of information that is received by, generated by, stored in, and transmitted from a device must be determined.[37] Some kinds of information may be regularly archived, either within an implanted neuroprosthetic device itself, within an external component of the system, or in some other external unit, facility, or system. Some information that is stored within a neuroprosthetic device may only be recoverable after the device is removed from its human host. For information that is encrypted, access to the cryptographic keys must be appropriately maintained.[38] If information is, for example, automatically archived wirelessly to some commercial cloud-based service, the InfoSec performance of that system and its interaction with the device must be carefully considered and tested.[39]

Some information that is generated or received by the system on an ongoing basis (e.g., input generated by a host's biological processes) might in effect be immediately 'destroyed' simply because it is used instantaneously by the system and not recorded in any format capable of being indefinitely preserved. Other information that has been recorded may be subject to intentional destruction. Complex legal, regulatory, and ethical considerations may dictate that some kinds of information must be preserved and other kinds may not be, especially in light of a neuroprosthetic device's status as a medical device that may gather sensitive health information about its user.[40]

---

[36] Such considerations would also apply when deploying countermeasures that are designed to directly affect the functioning of a system that is the source of an attack, if such a system itself may potentially be a neuroprosthesis that is part of a host-device system with a human host and if that person's psychological or biological functioning may be damaged or otherwise adversely affected by countermeasures enacted against the device. For a discussion of current legal and ethical issues involved, e.g., with the use of offensive countermeasures to mitigate a botnet, see Leder et al., "Proactive Botnet Countermeasures: An Offensive Approach" (2009).

[37] *NIST SP 800-100* (2006), p. 24.

[38] *NIST SP 800-100* (2006), p. 24.

[39] For information security issues specific to cloud-based systems, see Fernandes et al., "Security Issues in Cloud Environments: A Survey" (2013).

[40] Regarding regulatory and legal issues, see McCullagh et al.(2013); Patil & Turner (2008); and

2. Media sanitization

The sanitizing of storage media may involve processes such as overwriting existing data with new data, degaussing a magnetic drive, or physically destroying the media.[41] For those components of a neuroprosthetic system that exist outside of the human host's body, the sanitization of media may not raise any special considerations. However, the sanitization of media contained within the host's body must generally be performed in a way that does not harm the host. In some cases, it may be difficult or impossible to verify that information has been completely removed from the media without extracting the media from the host's body.

3. Hardware and software disposal

The disposal of neuroprosthetic hardware and software must be carried out in an appropriate manner in order to ensure information security.[42] The memory components of a used device may contain highly sensitive data about the cognitive and biological activity of its former human host. Moreover, the unit may be contaminated with biological matter that contains DNA from its former host and other substances that could be analyzed to reveal information about the host's health, activities, or environment.

III. Roles and responsibilities for information security

The security activities described above are carried out by one or more *participants* in an advanced neuroprosthetic system who fill specific *roles* that give them particular *responsibilities* for executing or overseeing security activities.[43]

A. Individuals and organizations participating in an advanced neuroprosthetic system

Typically, the production and use of an advanced neuroprosthetic system is a complex process that involves the participation of multiple individuals or organizations. Below we identify some of the common participants in that process and highlight ways in which they may contribute to the successful planning and execution of the InfoSec activities described above.

Kosta & Bowman (2012).

[41] *NIST SP 800-100* (2006), p. 24.

[42] *NIST SP 800-100* (2006), p. 24.

[43] See *NIST SP 800-100* (2006), p. 68.

1. Device designer

The organization or individual that designs a physical device may or may not be responsible for manufacturing it; the company that has developed the specifications and schematics for the device may then hire other firms to manufacture its components and subsystems or even to perform final assembly. The device designer may or may not also create the operating system that will run on the device (if it indeed utilizes an OS).[44]

Information security responsibilities

The designer of an advanced neuroprosthetic device plays the key role in determining those basic security controls that are built into the physical device at time of its manufacture: by determining the device's specifications and functionality, the designer also sets key physical and technological parameters that determine the kinds of security controls that can (or cannot) be implemented later by other parties and which determine the kinds of basic vulnerabilities, threats, and risks to which the device may be prone.

If a neuroprosthesis is produced with some security controls built into it as part of its original design specifications, the device's designer will likely exercise responsibility for security activities within the SDLC including carrying out a **determination of needs, security categorization, preliminary risk assessment, requirement analysis and development, risk assessment, cost consideration and reporting, security planning, security control development, developmental security test and evaluation,** and **other planning components,** insofar as they relate to the development and testing of those built-in controls. Note that because the designer may not have a clear understanding of the end purposes for which a device's operators will utilize the device, the controls developed by the designer may be of a basic and generalized nature and may need to be supplemented (or replaced) by other controls developed by the operator.

---

[44] Given the extent to which consumer electronics devices such as traditional desktop and laptop computers and mobile devices are subject to aftermarket hardware and software modifications by their purchasers and end users (even to the extent of the jailbreaking or rooting of smartphones by power users), it is impossible for a device's designer or manufacturer to know in advance exactly how end users might employ a product. (For that point, see Chadwick, "Therapy, Enhancement and Improvement" (2008).) It is likely that among 'power users' of advanced neuroprosthetic devices, a phenomenon (and even subculture) of modifying devices in unintended, unexpected, and potentially even illegal or unethical ways might emerge, as users push the limits of what is safe and possible for purposes of further enhancement and self-exploration. The ultimate hackers may not be those who hack external systems but who – through the use of neuroprosthetic devices – hack their own mind and brain. Denning et al. consider the possibility that while some users of neuroprostheses might modify their devices in unexpected and unapproved ways in order to "enhance their performance, increase their level of pain relief, or overstimulate the reward centers in the brain," others might do so intentionally to cause themselves harm; see Denning et al. (2009).

Once a device's design has been completed and manufacturing and installation of physical devices has begun, the designer's ability to physically modify the basic security controls built into the device may be limited or nonexistent.

2. Operating system developer

The organization or individual that develops the OS installed on a particular neuroprosthetic device may not have been responsible for the design of the device itself. For example, it may be the case (as is true now for many mobile devices utilizing the Android or Windows operating systems) that an existing operating system produced by one software designer is available for use by multiple device manufacturers, and manufacturers design their devices specifically to run that OS. Alternatively, a neuroprosthetic device that was distributed by its manufacturer with one OS installed could have that OS replaced by a different operating system that is manually installed by the device's user. On the other hand, it may be the case (as is true now for mobile devices utilizing the iOS operating system) that a single organization has designed both the physical device and the operating system with which it is distributed.

Information security responsibilities

The developer of an advanced neuroprosthetic device's operating system may incorporate some security controls directly into the OS. In this case, the OS developer will likely exercise responsibility for security activities within the SDLC including carrying out a **determination of needs, security categorization, preliminary risk assessment, requirement analysis and development, risk assessment, cost consideration and reporting, security planning, security control development, developmental security test and evaluation,** and **other planning components,** insofar as they relate to the development and testing of controls built into the OS.

By determining the basic framework and environment within which any other software or applications will run on the device, the developer of the OS sets key parameters that determine the kinds of security controls that can (or cannot) be implemented later by other parties and which determine the kinds of basic vulnerabilities, threats, and risks to which the device may be prone.

Because the developer of the OS may not have a clear understanding of the end purposes for which a device's operators will utilize the device, the controls developed by the OS developer may be of a basic and generalized nature and may need to be supplemented (or replaced) by other controls developed by the device's operator.

Once a device has been installed, it may or may not be possible for the developer of its OS to modify the basic security controls built into the operating system through software updates or upgrades. If the developer of an OS has the ability to access devices in use, gather information from the devices, and modify the OS, then the developer of the operating system may also exercise responsibilities in areas such as **configuration management and control**, **continuous monitoring**, and **information preservation**.

3. Manufacturer

When the word is used in a general sense without qualifications, it may be assumed that an organization that is described as the 'manufacturer' of an advanced neuroprosthesis has designed the physical device, designed or selected its operating system, physically produced the device by assembling its components, and installed the operating system. In a more specific sense, a 'component manufacturer' can be understood as an organization that produces one or more components that are incorporated into a device and the 'device manufacturer' can be understood as the organization that directs the process of assembling all components into a finished device. The device manufacturer may also have designed the device, or it may have produced the device on behalf of its designer.

Information security responsibilities

The manufacturer of an advanced neuroprosthetic device plays a key role in ensuring that the finished product that is ready for distribution, installation, and use possesses the security controls described in the design specifications for the device (and potentially also for its OS).

The manufacturer may exercise responsibility for security activities within the SDLC including carrying out **security control development**, **developmental security test and evaluation**, and **other planning components**, insofar as they relate to the development and testing of controls built into the device and its OS. The manufacturer may also have responsibility for **hardware and software disposal** for components and finished units that are produced but not distributed or which are returned by the device's users due to manufacturing defects, the termination of a leasing arrangement, or other circumstances.

4. Provider

A device's provider is the organization that makes it available to the user for implantation and use. In some cases, a device's manufacturer may be a company that directly sells or leases the device to users; in that case, the manufacturer is also a provider. In the case of an advanced neuroprosthesis that is used for therapeutic purposes, a hospital, charitable organization, insurance company, or government agency may provide the device to the human host who is also its user, either for a fee or free of charge. In the case of an advanced neuroprosthesis that is implanted into soldiers for military use, the military organization may retain ownership and control of the device and serve as its operator while it is implanted in its human host. In this situation, one military agency that has produced the device may have provided it to another military agency that serves as the end user; in the sense in which we use the word here, the device has been physically implanted in its human host but not 'provided' to the soldier for his or her use, as he or she may or may not even realize that the device has been implanted.[45]

Information security responsibilities

The provider of an advanced neuroprosthetic device may or may not play specific roles in information security within the device's SDLC. At a minimum, the provider will ensure adequate information security for devices from the time when they are received from the manufacturer to the time of their delivery to the operator for implantation in the host.

Some providers may acquire neuroprosthetic devices from manufacturers and add their own customized components, operating system, or other software before delivering them to operators; in this case, a device's provider may exercise responsibility for security activities within the SDLC for their customized system including carrying out a **determination of needs, security categorization, preliminary risk assessment, requirement analysis and development, risk assessment, cost consideration and reporting, security planning, security control development, developmental security test and evaluation,** and **other planning components**, insofar as they relate to modification of the device from its original factory-default condition. The provider may also have responsibility for **hardware and software disposal** for units that are received from the manufacturer but never delivered to operators or which are returned by the devices' users due to manufacturing defects, the termination of a leasing arrangement, or other circumstances.

5. Installer / implanter

The installer physically integrates a neuroprosthetic device into the physical organism and neural circuitry of its human host and activates the device. For some advanced neuroprosthetic devices, this will involve surgical implantation that is performed in a specialized medical facility. In other cases, the human host may already have been surgically implanted with a prosthetic sock, port, or jack into which neuroprosthetic devices (potentially of many different kinds) can easily be inserted; in such a situation, it may be possible for the installation of a particular neuroprosthetic device to be performed by

---

[45] For the possibility that human hosts might unwittingly be implanted, e.g., with RFID devices, see Gasson, "Human ICT Implants: From Restorative Application to Human Enhancement" (2012).

the device's host or another individual without any special training and out-side of a medical facility.

Information security responsibilities

The installer who performs implantation of an advanced neuroprosthetic device into its human host and activation of the device plays a key role in ensuring that the security controls built into the device and software are able to function successfully. Failure to implant the device correctly or to fully ac-tivate its relevant features may create vulnerabilities, threats, and risks that compromise the information security of its operator and host.

The installer of an advanced neuroprosthetic device may bear particular responsibility for security activities such as the final pre-implementation **se-curity test and evaluation** of the particular unit to be implanted, **inspection and acceptance** of the unit to be implanted, **system integration and installation**, **secu-rity certification** of the particular unit implanted, **configuration management and control**, and **continuous monitoring**. If the installer also participates in the repair or removal of neuroprosthetic devices that have been implanted in a human host, he or she may also have responsibility for **information preservation**, **media sanitization**, and **hardware and software disposal**.

6. Application developer

For some neuroprosthetic devices, the only software that a devices are ca-pable of running is the operating system installed by the device's manufac-turer; there may not even be any physical means of accessing the device's memory to install new software. For other neuroprostheses, it may be possi-ble to install particular applications on a device and to run that software – regardless of whether or not this possibility was foreseen and intended by the device's designer and manufacturer.

Application developers may include parties from the device's designer and manufacturer to authorized providers of third-party applications, to the de-vice's operator or host who create their own customized apps,[46] to hackers who create specialized software which once installed allows them to directly analyze and control the device, to cybercriminals who create computer worms or other malware that can autonomously infect the unit.

Information security responsibilities

The developer of an application for advanced neuroprosthetic device plays a role in ensuring information security for devices on which that software is installed and run. Depending on the circumstances, the application developer

---

[46] For factors that might cause a device's host to (licitly or illicitly) alter the functionality of his or her own implanted device, see Denning et al., "Patients, pacemakers, and implantable defib-rillators: Human values and security for wireless implantable medical devices" (2010).

may or may not even realize that his or her application will be run on neuro-prosthetic devices; if the application is designed to run on a general-purpose operating system such as Windows or Android, then neuroprosthetic devices utilizing the relevant operating system may be able to run the application, regardless of whether this was envisioned by its developer.

Software developers may exercise responsibility for security activities within the SDLC relating to their applications, including carrying out a **determination of needs, security categorization, preliminary risk assessment, requirement analysis and development, risk assessment, cost consideration and reporting, security planning, security control development, developmental security test and evaluation,** and **other planning components.** If the software developers have an official relationship with the device's designer, OS developer, manufacturer, provider, or operator, the software developer may carry out security activities in close collaboration with those partners and with a high level of diligence. Some software developers (including individual programmers who are not affiliated with a large company) may not have the resources needed to conduct some of the security activities at a robust level.

After an application has been delivered to a device's operator for installation on the device, the application's developer may or may not subsequently have an opportunity to modify the software and its security controls through software updates or upgrades. If the developer of an application has the ability to access devices in use that have the app running or installed, gather information from the devices, and modify the app, then the developer of the application may also exercise responsibilities in areas such as **configuration management and control, continuous monitoring,** and **information preservation** – either for the developer's own internal purposes or at the specific request of the device's operator (e.g., if the software developer maintains a cloud-based platform that services their application).

7. Application provider

In some cases, it may be possible for an advanced neuroprosthetic device's operator or host to acquire third-party software directly from its developer and to install it on the device without any involvement or oversight on the part of the device's designer, manufacturer, or provider. In other cases, the device's designer, manufacturer, or provider may maintain control over the software that can be installed on the device by creating administrative, logical, or physical controls that require all software to be selected and installed using a proprietary mechanism (such as a cloud-based app store) that is controlled by that party.

Information security responsibilities

The application provider for an advanced neuroprosthesis plays an important role in ensuring information security for the device by controlling the kinds of applications that can be installed and run on the device. Some 'application providers' may simply be automated systems that allow any third-party developer to upload applications into the app store for download by users without any particular scrutiny being undertaken of the software's functionality or InfoSec characteristics. In other cases, the application provider may serve as the developer of all of the software that it makes available for installation, or it may scrutinize software submitted for approval by third-party developers by carrying out security activities such as **risk assessment** and **security planning** to determine the criteria that must be met by third-party applications, a **security test and evaluation** and **inspection and acceptance** prior to making the software available for use by a device's operators, **system integration and installation** of software not into the neuroprosthetic devices themselves but into the provider's distribution systems, potentially a process of **security certification** to reassure device operators, **configuration management and control** to ensure compatibility and interoperability with other installed software, **continuous monitoring** of general performance and InfoSec controls, and potentially **media sanitization** and **hardware and software disposal** mechanisms that ensure information security upon uninstallation of the software by a device's operator.

8. Operator / user

The operator or user of an advanced neuroprosthetic device is the organization or individual who controls the device's functioning. This may or may not also be the device's human host or the primary beneficiary of its functioning.[47] In the case of a neuroprosthesis whose functioning is directly controlled by the thoughts of its human host, the host would also be the device's operator.

In the case, for example, of an artificial eye that can record and transmit live video and which has been implanted into an agent by a government organization for military or intelligence-gathering purposes, specialized government personnel may maintain ongoing remote control over the device and direct its functioning; in this case, they would be the device's operators, and the device's human host may or may not even realize the purposes for which the device is being employed. In some cases, a device's host may be able to control some aspects of a device's functioning while medical or technological specialists control other aspects; in those situations, both parties serve as us-

---

[47] See Chapter Three of this text for a discussion of the distinction between a device's host and user.

ers and operators of the device. In other cases, a device may have no mechanism for receiving instructions from its host or any other external agent but is instead directed entirely by its own software or internal artificial intelligence; in such a situation one could say that the device has no operator (or is its own operator).

Information security responsibilities

The operator of an advanced neuroprosthesis plays a critical role in ensuring information security by exercising responsibilities for security activities such as carrying out a **determination of needs, security categorization, preliminary risk assessment, requirement analysis and development, risk assessment, cost considerations and reporting**, and **security planning** for the specific environment within which and purpose for which the device will be deployed; **security control development** to design controls that are relevant to the specific circumstances in which the device will be used and which supplement or replace those basic controls built into the device's physical components, OS, and applications; **developmental security test and evaluation; other planning components** (including preparing contracts and agreements with the provider, installer, and other organizations or individuals that may provide maintenance services, as well as with the device's human host); a final pre-deployment **security test and evaluation** of the system as initially configured; **inspection and acceptance** of the device as initially configured; **system integration and installation** of the device to integrate it with its host's neural circuity (through physical implantation in the host's body or through other means); and potentially **security certification** and **security accreditation** for the device in its intended environment and use.

The operator will especially bear responsibility for security activities including **configuration management** and **control and continuous monitoring** of the device after its installation and activation. The operator may also play the lead role in in **information preservation** during and after the device's period of active use and **media sanitization** and **hardware and software disposal** upon completion of its service lifetime.

9. Human host

The host of an advanced neuroprosthesis is the human being into whose neural circuitry the device has been integrated. In some cases (e.g., with some forms of motor neuroprostheses), an advanced neuroprosthetic device could potentially operate externally to the host's body in a noninvasive fashion,[48]

---

[48] See Panoulas et al., "Brain-Computer Interface (BCI): Types, Processing Perspectives and Applications" (2010); Lebedev, "Brain-Machine Interfaces: An Overview" (2014); Birbaumer & Haagen, "Restoration of Movement and Thought from Neuroelectric and Metabolic Brain Activity: Brain-Computer Interfaces (BCIs)" (2007); and Gerhardt & Tresco, "Sensor Technology" (2008).

although this raises questions of the extent to which the device has truly been 'integrated into' the host's neural circuitry.[49]

In this text it is presumed that neuroprosthetic devices that are fully integrated into their host's neural circuitry typically involve implantation into the host's biological organism. (In some cases this may involve a stable long-term interface, although models of future neuroprosthetic systems comprising, for example, swarms of nanorobots operating within the host's brain have also been proposed.[50]) However, there may be situations in which an initial neuroprosthetic device has been installed that includes a standardized external socket, port, or jack through which supplementary components or even entirely separate neuroprosthetic devices or external systems can be installed; in those cases, new newly-installed devices become integrated with the host's neural circuitry but without necessarily enjoying any direct interface with the host's natural biological structures or processes. Instead, the newly attached device interfaces directly with the existing neuroprosthetic socket and, through it, indirectly with the host's biological neural systems.

A neuroprosthetic device's human host may have purchased or leased the device, may have received it as a therapeutic aid or personal gift, may have received it as a productivity tool or fringe benefit from an employer, or may potentially have created the device himself or herself. The device may be implanted either permanently or for a predetermined or indeterminate temporary period. The human host may or may not be able to directly control the device and may or may not even realize that the device has been implanted.

Information security responsibilities

The host of an advanced neuroprosthetic device may or may not play a particular role in carrying out security activities designed to ensure its information security. In some cases, the host may not even realize that the device has been implanted in his or her body or (e.g., in the case of a host who is in a coma) may not be able to intentionally undertake any actions in support of the system's information security. In other cases, the device's host may also be its operator and directly control its use.

In his or her role as the human host of an advanced neuroprosthesis, a person may especially bear responsibility for participating in security activities such as **inspection and acceptance** of the device, **system integration and installation** through implantation of the device and activation of its features, **security certification** of the implanted device, **configuration management and control**, **continuous monitoring**, and **hardware and software disposal** upon the conclusion of a system's service lifetime.

---

[49] See Chapter One of this text for a discussion of such questions.
[50] See Al-Hudhud, "On Swarming Medical Nanorobots" (2012).

10. Information security collaboration among participants

In the context of a large institution such as a military agency, an organization can potentially ensure the coherence and effectiveness of InfoSec efforts by unifying the roles of device designer, OS developer, manufacturer, provider, installer, application developer, application provider, and operator of an advanced neuroprosthesis, while the device's human host may also belong to the organization.

On the other hand, in the case of a neuroprosthetic device intended for broad use as a consumer electronics product, the situation can be very different: collectively, there is still a need to provide information security for such devices, but the system responsible for ensuring that security may not be a single unified organization but rather a disjointed patchwork of organizations and individuals who have very different interests, motives, and capacities and who may not have any ongoing relationship with one another. One participant in the system may or may not be able to assume good will and competence on the part of all the other parties. This can create significant obstacles to the pursuit of information security.

B. Typical formalization of roles and responsibilities in an institutional setting

Within a government agency or large corporation the personnel structure for information security generally has several levels. Though the roster of organizational roles be more complex,[51] such a structure often includes the key roles of **chief information officer** (CIO), **information system owner**, **information owner**, and **information system security officer** (ISSO).

At the most general level of policy are individuals like the CIO whose task is to ensure that information security efforts are consistent with the organization's broad strategic goals and priorities and to allocate resources among and set objectives for his or her staff. At the most concrete levels are expert personnel with particular specializations (such as ISSOs) who work daily to implement InfoSec policies effectively for specific systems. Below we consider the responsibilities traditionally associated with these roles and the ways in which they might relate to advanced neuroprosthetic systems used within an organization.

---

[51] For example, a government agency may also have a senior agency information security officer (SAISO) who serves as a liaison between the CIO and ISSOs. See *NIST SP 800-100* (2006), p. 70.

1. Chief information officer (CIO)

The CIO is the senior individual responsible for developing and maintaining information system security throughout an organization.[52] He or she develops job descriptions for key InfoSec positions and ensures that highly qualified individuals are recruited, successfully trained and integrated into the organization, and supported in their work. Depending on the size of the organization, the degree to which the CIO is personally involved with developing all of the organization's information security policies, practices, and mechanisms will vary; however, he or she does bear ultimate responsibility for ensuring that such procedures are developed and successfully implemented.

The CIO collaborates with leaders of other departments within the organization to ensure that the InfoSec priorities and activities are consistent with and support the fulfillment of the organization's legal and regulatory obligations, ethical aims, and strategic and financial objectives. Conversely, the CIO ensures that the expertise, insights, and objectives of personnel within the InfoSec office inform and shape, as appropriate, the overall strategic direction and decisions of the organization as well as the work of other individual departments.

With regard to ensuring information security for neuroprosthetic devices and systems, the CIO has potential responsibilities relating to at least four different spheres: 1) the use of neuroprostheses by the organization's **customers and clients**; 2) the use of neuroprostheses by the organization's **employees and personnel**; 3) the use of neuroprostheses by the organization's **competitors**; and 4) the use of neuroprostheses by **hackers, cybercriminals, and other outside parties** who may potentially be employed by the company's competitors as part of business intelligence or corporate espionage campaigns or who may have their own financial, political, or personal motives.

Responsibilities regarding neuroprostheses used by the organization's customers and clients

For a company that produces neuroprosthetic technologies, it would be individuals such as the chief technology officer (CTO) or chief marketing officer (CMO) – and not the CIO – who is primarily responsible for envisioning and developing new products and ensuring that they will provide InfoSec capacities and features that are satisfactory for end users.

If, however, the company provides ongoing monitoring and maintenance for its devices, maintains a centralized 'app store' that allows users to acquire new software for their devices, provides cloud-based storage for use by the devices, or otherwise integrates the neuroprostheses that it manufactures

---

[52] *NIST SP 800-100* (2006), p. 68.

into the company's technological systems after the devices are in use by consumers, the CIO would play the key role in ensuring the information security of such systems and arrangements.

Even for an organization that has no involvement with the production or distribution of neuroprosthetic devices, the organization's CIO must be concerned with ways in which the use of neuroprostheses by customers may impact the company's information security. For example, some organizations that produce films, live theater, art exhibitions, concerts, or sporting events have developed (and attempt to enforce) policies that bar customers from videotaping such events and, in particular, from making any such unauthorized videos publically accessible (e.g., through a video streaming site). Such policies may become legally and ethically untenable – not to mention commercially unsound – if in the future sufficient numbers of human beings possess artificial eyes that continually record live video of everything that their hosts visually experience and the streaming and public sharing of such videos comes to be seen as a normal part of human existence – and potentially even a protected form of freedom of expression and the 'fair use' of others' intellectual property. In such a situation, questions regarding the protection of an organization's proprietary information and intellectual property may involve complex legal and technological questions that a CIO would need to address.

Responsibilities regarding neuroprostheses used by the organization's employees and personnel

The use of advanced neuroprostheses by an organization's personnel may take one of two forms:

- **Neuroprostheses personally acquired and owned by an organization's personnel.** It is possible that individuals who happen to work in the organization may have acquired such devices on their own, for their own personal ends and using their own financial resources. In this case, the organization may not necessarily even know whether its personnel possess such devices – and may even be legally barred from attempting to discover whether they do, insofar as an individual's status as someone who does or does not possess neuroprosthetic devices may be legally protected personal health information, and making employment-related decisions on the basis of such information could, in some circumstances, be considered a form of illegal employment discrimination. A CIO may thus need to ensure that his or her organization's security controls are robust enough to deal with the case, for example, of an employee who possesses an artificial eye and who (without any ill intentions on the employee's part, and perhaps even without the employee's knowledge) is recording everything that appears on the employee's computer screen – without the organization being able to know whether any of its personnel actually possess such devices.

At some point, the organization may choose to proactively take advantage of the fact that some of its personnel not only voluntarily *acknowledge* the fact that they have acquired neuroprosthetic devices but actively request that they be allowed to interconnect their devices with the organization's information systems in order to more effectively and efficiently carry out their work. Such a phenomenon would follow the pattern seen in recent decades with mobile devices and the gradual development of 'bring your own device' (BYOD) policies in which organizations attempt to support the use of official organizational systems (such as email, CRM and HRM software, and countless other cloud-based systems) on the bewildering array of mobile devices that have been personally acquired by the organization's personnel.[53]

- **Neuroprostheses provided by an organization to its personnel.** There may be situations in which an organization develops or acquires neuroprosthetic devices that it then implants in its personnel. In this case, the organization may – subject to relevant laws and employment agreements with its personnel – maintain ownership of the devices even after they are implanted and may bear ultimate responsibility for the information security of the devices, which may be fully and directly integrated into (and even controlled by) organizational information systems. Such a situation might occur, for example, with a military organization that provides its personnel with advanced neuroprostheses intended for use in combat situations, intelligence gathering, or other operations.

Responsibilities regarding neuroprostheses used by the organization's competitors

If an organization's competitors are utilizing neuroprosthetic devices in order to work more effectively and efficiently, develop new products and services to offer to customers, or otherwise achieve competitive advantages, it would likely be senior executives such as the organization's Chief Human Resources Officer (CHRO), CMO, or Chief Strategy Officer (CSO) – and not the CIO – who lead the debate about whether the organization should itself utilize advanced neuroprostheses among its personnel in order to build or maintain a competitive advantage.

---

[53] See Armando et al., "Formal Modeling and Automatic Enforcement of Bring Your Own Device Policies" (2014), for a discussion of some issues relating to BYOD issues and proposed solutions.

On the other hand, if an organization's competitors are employing advanced neuroprosthetic devices in an effort to conduct business intelligence operations against the organization or otherwise acquire sensitive or proprietary information that would be of use to the competitors in undermining the organization's competitive position or advancing their own interests, the organization's CIO would play a key role in ensuring that the organization's InfoSec policies, practices, and mechanisms minimize competitors' ability to utilize neuroprosthetic devices to acquire such information. Competitors' efforts to acquire business intelligence about the organization may utilize legitimate and accepted techniques that are neither illegal nor unethical, and the extent to which an organization can attempt to block the gathering of such intelligence may be constrained by legal, ethical, and cultural factors. However, the CIO will attempt to ensure that – while respecting such constraints – the organization's information security is not compromised by competitors' use of neuroprosthetic devices to gather information.

Responsibilities regarding neuroprostheses used by hackers, cybercriminals, and other outside parties

An organization's CIO takes the lead in protecting the organization's information against the dangers posed by individuals or organizations that are willing to employ potentially illegal or unethical means in order to gain unauthorized access to the organization's systems and information. Such efforts at gaining unauthorized access may sometimes be undertaken by hackers who are trying to break into a system simply in order to prove to themselves that it can be done and to discover how the system works but who have no desire to otherwise harm the organization or financially exploit such access.[54] On the other hand, efforts at gaining unauthorized access to a system and its information may sometimes be made by cybercriminals who are intent on exploiting such access for purposes of blackmail, extortion, or otherwise harming the company (and potentially its individual employees and customers).

Insofar as advanced neuroprosthetic devices provide their users with enhanced sensory, cognitive, or motor capacities and allow them to interface with physical and virtual systems in ways that were previously impossible, hackers and cybercriminals may develop techniques for utilizing such neuro-

---

[54] For one discussion of hackers' motives and interests, see Zaród, "Constructing Hackers. Professional Biographies of Polish Hackers" (2015).

prostheses in ways that facilitate their ability to gain and maintain unauthorized access to systems and information.[55] The potential use of advanced neuroprostheses by unauthorized parties to gain access to an organization's systems and information creates at least two issues that a CIO must consider:

- **The creation of new kinds of vulnerabilities, threats, and risks.** The use of such neuroprostheses may create entirely new kinds of vulnerabilities, threats, and risks that did not previously exist and which the organization must address through its InfoSec planning and procedures.

- **The risk of harming human beings through the application of information security countermeasures.** A CIO must take into account the fact that some kinds of proactive or offensive countermeasures that an organization can theoretically deploy against potential or actual attacks have the capacity to disrupt the performance, functioning, and, in principle, even structural components of the remote systems responsible for initiating and undertaking the attack. Complex legal and ethical questions already surround the use of proactive countermeasures to repel or disrupt cyberattacks (e.g., launching temporary denial of service attacks against command-and-control servers in an effort to mitigate botnets[56]), and these questions become even more complex – and potentially create new legal and moral responsibilities for an organization – if the remote computer that is launching a hacking attempt and which might be disrupted or damaged by the organization's countermeasures is not a conventional desktop computer or server but an advanced neuroprosthetic device implanted in the brain of a human being – who may be the hacker who is directing the attack or may simply be an innocent third party whose neural implant has been hijacked and, unbeknownst to the person, is being used to participate in the attack. In this case, disrupting the operations of that neuroprosthetic device or otherwise damaging it could result in significant temporary or permanent psychological or physical harm – potentially including even death – for the host in whom the device is implanted, if the device possesses interconnections with the host's biological and cognitive systems that make those systems dependent on the device's correct functioning or subject to damage by its anomalous behavior. This raises complicated questions regarding legal, ethical, and financial liability on the part of an organization that deploys such countermeasures.

---

[55] See Chapter Three of this text for a discussion of possible uses of neuroprosthetic devices as a means for carrying out such attacks.

[56] See Leder et al. (2009).

2. Information system owner

The information system owner is "responsible for the overall procurement, development, integration, modification, and operation and maintenance of the information system."[57] This individual participates in the development of the system security plan, and upon implementation of the system he or she ensures that personnel using the system receive the relevant information security training and use the system in a way consistent with the system security plan.[58]

3. Information owner

The information owner is the individual who has been given "authority for specified information and is responsible for establishing the controls for information generation, collection, processing, dissemination, and disposal."[59] Based on his or her expert knowledge of the relevant legal, ethical, and operational considerations, the information owner determines which individuals or systems should have which levels of access to which kinds of information and the ways in which they should be allowed to use or manipulate the information. He or she works closely with a system's designers and operators to ensure that the security controls developed and implemented for the system provide the appropriate levels of access to the relevant individuals or systems.

In the case of an advanced neuroprosthetic device, complex legal and regulatory regimes govern the ownership of information that is received, created, stored, or transmitted by the device.[60] Depending on the exact nature of the device, the kind of information that it handles, and the contracts and agreements that have been formed between the relevant parties, information may be owned by the human host in whom a device is implanted, the device's manufacturer, the device's operator, or unrelated outside parties. For example, a single neuroprosthetic device might contain data about the biological processes of its host that has been gathered by the device and which is considered personal health information; proprietary operating system software that was created and installed by the firm that manufactured the device; proprietary application software that was created and installed by the host's em-

---

[57] *NIST SP 800-100* (2006), p. 69.

[58] *NIST SP 800-100* (2006), p. 69.

[59] *NIST SP 800-100* (2006), p. 69.

[60] See Kosta & Bowman (2012); McGee, "Bioelectronics and Implanted Devices" (2008); Mak, "Ethical Values for E-Society: Information, Security and Privacy" (2010); McGrath & Scanaill, "Regulations and Standards: Considerations for Sensor Technologies" (2013); and Shoniregun et al., "Introduction to E-Healthcare Information Security" (2010).

ployer that is operating the device; and third-party text, video, audio, or image files that have been downloaded from the Internet onto the device by its user or host.[61]

4. Information system security officer (ISSO)

In contrast to the CIO, who has responsibility for all aspects of information security for the entire organization, an information system security officer (ISSO) focuses on ensuring information security for one or more particular systems within the organization. The ISSO's focus on information security for the system also contrasts with the role of the information system owner, who must manage a broader range of strategic, legal, financial, and operational aspects of the information system that may or may not directly relate to its information security. The ISSO's role also differs from that of the information owner, insofar as the information owner is concerned first and foremost with the information as such (rather than the system or systems that are currently used to make it accessible to users) and he or she may lack expertise in the technological and security aspects of the information system that the ISSO should possess. The ISSO participates in the development of the system security plan and implementation of security controls and coordinates the work of the relevant individuals to ensure that the system security plan and security controls are updated as needed to respond to any future changes that must be made to the information system for operational or other reasons.[62]

In the case of an advanced neuroprosthetic device, an ISSO would need to combine expertise in in the field of information security with knowledge of fields such as biology, psychology, medicine, and biomedical engineering.

C. Potential specialized roles for future neuroprosthetic systems

An advanced neuroprosthetic system will typically include the kinds of participants described above, such as a device designer, manufacturer, operator, and human host; if the manufacturer and operator are large institutions, they may also include individuals who have been formally assigned roles such as those of chief information officer (CIO) or information system security officer (ISSO).

Depending on the exact form, capacities, environmental context, and intended use that a particular neuroprosthetic system possesses, some of the individuals involved with developing, producing, maintaining, operating, and securing that system may possess specialized roles that are not relevant for

---

[61] See the device ontology in Chapter One of Gladden (2017) for an overview of the different kinds of possible information.

[62] *NIST SP 800-100* (2006), p. 70.

other kinds of neuroprosthetic devices or information systems more broadly. Such roles may be formally assigned or informally adopted. Examples of such roles that may be relevant for certain kinds of future neuroprosthetic devices and whose successful performance has implications for the information security of the neuroprosthetic system in which the individuals filling these roles participate include:

- **Body schema engineer / body designer.** A body schema engineer participates in the development, production, installation, or operation of advanced neuroprosthetic systems that provide their user with the experience of possessing and utilizing a body that is (potentially radically) nonhuman.[63] This would occur, for example, if a human being uses a neuroprosthetic device that replaces his or her natural sensory input with sensory input that creates the experience of possessing a nonhuman body that is inhabiting and operating within some virtual environment. Such possibilities already exist, for example, through the use of virtual reality headsets and haptic feedback with first-person video games in which the player takes on the role of a nonhuman character. Alternatively, the physical body of a human being (apart from the brain) could be replaced with an artificial cybernetic body that is nonhuman in its form and functioning.[64] A body schema engineer would ensure that the human being's mind was capable of adapting successfully to the new nonhuman body and the potentially radically different biocybernetic sensorimotor feedback loop that it creates.

  Many unique issues relating to information security would result. For example, the human owner of a new nonhuman body might discover that particular kinds of thoughts or volitions which – in the person's human body – had created purely internal mental phenomena will in this new body create gestures, expressions, vocalizations, or other physical expressions that can be detected and interpreted by others in a way that conveys information that the user would like to keep confidential.

- **Neuromnemonic engineer / mnemonic designer / memory hacker.** Some kinds of advanced mnemoprosthetic devices may affect or participate in a human mind's processes of encoding, storing, or retrieving memories in a way that allows the intentional alteration, enhancement, or removal of existing memories or the creation of new memories. Such

---

[63] See Gladden, "Cybershells, Shapeshifting, and Neuroprosthetics: Video Games as Tools for Posthuman 'Body Schema (Re)Engineering'" (2015).

[64] See Gladden, "Cybershells, Shapeshifting, and Neuroprosthetics" (2015).

future technologies allowing for the editing of human memories may build on experimental results that have already been achieved with the use of optogenetic neuromodulation systems for the editing of memories in mice.[65]

A neuromnemonic engineer would develop tools and techniques for creating particular kinds of mnemonic content or bringing about particular kinds of targeted changes in existing mnemonic content within a human host's memory systems, while a mnemonic designer would utilize those tools to fashion a particular memory or 'remembered experience' for some specific purpose. Significant questions of information security arise, insofar as the application of neuromnemonic engineering tools and techniques could, if used improperly, inadvertently destroy the only copy of information existing within a person's mind, and it may in some circumstances even be used intentionally to destroy particular information.[66] Moreover, the use of mnemocybernetic technologies to affect a particular person's memories (or even a human being's awareness of the possibility that such technologies *may* have been used) may create uncertainty for that individual about whether the information stored in his or her memories is true or false and an inability to trust any of his or her apparent memories.[67] The use of such technologies may also provide a mnemonic designer with the ability to access, interpret, and create an external record of (and perhaps even potentially 'export' in an automated fashion) memories stored within the human host's mind that he or she wishes to keep confidential.

The extent to which mnemocybernetic technologies and neuromnemonic engineering can someday be implemented successfully for human beings is still unclear and will depend on the resolution to numerous outstanding questions in the field of neuroscience. For example, if holographic memory models (like the Holonomic Brain Theory[68]) are correct, the ability to create, modify, or delete complex, content-rich memories through the manipulation of a small number

---

[65] See, e.g., Han et al., "Selective Erasure of a Fear Memory" (2009), and Ramirez et al., "Creating a False Memory in the Hippocampus" (2013).

[66] For the possibility that an adversary might use a compromised neuroprosthetic device in order to alter, disrupt, or manipulate the memories of its host, see Denning et al. (2009).

[67] See Gladden, "Neural Implants as Gateways to Digital-Physical Ecosystems and Posthuman Socioeconomic Interaction" (2016).

[68] See, e.g., Longuet-Higgins, "Holographic Model of Temporal Recall" (1968); Pribram, "Prolegomenon for a Holonomic Brain Theory" (1990); and Pribram & Meade, "Conscious Awareness: Processing in the Synaptodendritic Web – The Correlation of Neuron Density with Brain Size" (1999).

of neurons may prove not simply practically difficult but theoretically impossible.

- **Sensory engineer / sense sculptor / sense hacker.** Even if the basic shape and structure of a human being's body are not altered by a neuro-prosthetic device, the device may subtly or radically reshape the way in which the body is used to experience the external environment. For example, a human being who appears completely normal but who possesses relevant neuroprostheses may be experiencing the world by seeing with infrared and telescopic vision, hearing ultrasonic phenomena, and receiving other sense data generated by magnetic fields, airborne chemicals, or other substances or phenomena in the environment that natural unaugmented human beings are unable to detect.[69] Moreover, it is theoretically possible for sense data acquired through particular sensory organs to be 'remapped' or 'rerouted' so that it is received by the brain and perceived and interpreted by the mind using nonstandard sensory modalities. For example, instead of hearing ultrasound, it might be possible to 'see' it; it may also be possible to directly 'see' the existence of magnetic fields or to 'hear' the presence of certain chemicals in the atmosphere.[70] A sensory engineer would develop technologies to allow for the remapping of sense data and alteration and enhancement of sensory modalities, while a sense designer or 'sense sculptor' would work to fine-tune a person's sensory modalities to create the capacity for unique sensory experiences that are artistically and aesthetically meaningful and potentially unique. Questions of information security arise in such work, insofar as sense data can potentially be altered or 'enhanced' in such ways that it no longer conveys accurate or useful information about the external environment;[71] moreover the technologies used to effect such sensory enhancement may potentially create vulnerabilities that could allow unauthorized parties to access a host's sensory experiences by accessing the devices that present them to the host's mind.

D. Multiple roles for a single individual?

A single individual may fill more than one organizational InfoSec role, either temporarily or permanently. However, whenever a single person fills multiple roles there is a danger that the effectiveness of the security planning process may be undermined: having each role filled by a different person may

---

[69] See Warwick, "The Cyborg Revolution" (2014); Gasson et al., "Human ICT Implants: From Invasive to Pervasive" (2012); and Merkel et al., "Central Neural Prostheses" (2007).

[70] See Wiener, *Cybernetics: Or Control and Communication in the Animal and the Machine* (1961), loc. 2784ff; Lebedev (2014), p. 106; Warwick (2014); and Chapter Two of Gladden (2017).

[71] See Chapters Two and Three of Gladden (2017) for a discussion of authenticity in sensation.

increase the likelihood that the unique aims and interests associated with a role will be given the attention that they require and will be vigorously pursued as the individuals filling different roles negotiate a common security plan based on their different (and sometimes potentially even competing) interests. If one person fills multiple roles, the negotiation and planning process can be short-circuited by this conflict of interest as the person decides to privilege one of his or her roles above the other(s).[72]

If one individual fills multiple roles during the operations and maintenance phase of the SDLC, this can also cause problems insofar as the effectiveness of some administrative, logical, and physical controls may be grounded in the fact that each individual within the organization is expected to fill only a single role, and disabling or bypassing the control would require collusion among two or more individuals; an individual who is simultaneously filling multiple roles may (if only temporarily) be able to singlehandedly bypass the control and compromise the information security of the entire system.

In the case of an advanced neuroprosthetic system, complications may arise from the fact that typically the role of serving as a device's human host is permanently filled by a single person – who may also be the device's sole operator and who has an intense interest in the safety and security of the device. In such circumstances, the host may fill multiple roles simultaneously, and he or she must be adequately trained and educated to understand the different aims and interests connected with different InfoSec roles and the need to prudently balance the claims of competing interests when performing activities relating to the different roles.

E. Roles and responsibilities for nonhuman agents?

While technological systems play key roles in the performance of many security activities, it is generally the human being who oversees the operation of that system who is said to possess responsibility for the activity and thus to fill a particular role. The extent to which legal and ethical responsibility for the performance of certain tasks could be attributed to a technological system such as an instantiation of artificially intelligent software, a social robot, or a sapient robotic network or swarm is a matter of ongoing debate.[73]

---

[72] *NIST SP 800-100* (2006), p. 68.

[73] For discussion of such questions in various contexts, see, e.g., Coeckelbergh, "From Killer Machines to Doctrines and Swarms, or Why Ethics of Military Robotics Is Not (Necessarily) About Robots" (2011); Calverley, "Imagining a non-biological machine as a legal person" (2008); Datteri, "Predicting the Long-Term Effects of Human-Robot Interaction: A Reflection on Responsibility in Medical Robotics" (2013); Hellström, "On the Moral Responsibility of Military Robots" (2013);

IV. Conclusion

As we have seen in this chapter, the work of ensuring information security for a neuroprosthetic system often involves collaboration among many individuals who fill different roles that give them responsibility for carrying out particular InfoSec activities. In the following three chapters, we will explore in greater detail the nature and importance of many of these activities, which can be grouped into three main categories as preventive, detective, and corrective or compensating controls.

Kuflik, "Computers in Control: Rational Transfer of Authority or Irresponsible Abdication of Autonomy?" (1999); Kirkpatrick, "Legal Issues with Robots" (2013); Stahl, "Responsible Computers? A Case for Ascribing Quasi-Responsibility to Computers Independent of Personhood or Agency" (2006); Weber & Weber, "General Approaches for a Legal Framework" (2010); and Gladden, "The Diffuse Intelligent Other: An Ontology of Nonlocalizable Robots as Moral and Legal Actors" (2016).

Chapter Six

Preventive Security Controls for Neuroprosthetic Devices and Information Systems

Abstract. This chapter explores the way in which standard preventive security controls (such as those described in *NIST Special Publication 800-53*) become more important, less relevant, or significantly altered in nature when applied to ensuring the information security of advanced neuroprosthetic devices and host-device systems. Controls are addressed using an SDLC framework whose stages are (1) supersystem planning; (2) device design and manufacture; (3) device deployment; (4) device operation; and (5) device disconnection, removal, and disposal.

Preventive controls considered include those relating to security planning; risk assessment and formulation of security requirements; personnel controls; information system architecture; device design principles; memory-related controls; cryptographic protections; device power and shutoff mechanisms; program execution protections; input controls; logical access control architecture; authentication mechanisms; session controls; wireless and remote-access protections; backup capabilities; component protections; controls on external developers and suppliers; environmental protections; contingency planning; system component inventory; selection of device recipients and authorization of access; physical and logical hardening of the host-device system and supersystem; device initialization and configuration controls; account management; security awareness training; vulnerability analysis; operations security (OPSEC); control of device connections; media protections; exfiltration protections; maintenance; security alerts; information retention; and media sanitization.

Introduction

In this chapter, we review a wide range of standard preventive security controls for information systems and identify unique complications and situations that arise from the perspective of information security, biomedical engineering, organizational management, and ethics when such controls are applied to neuroprosthetic devices and larger information systems that include neuroprosthetic components. The text provides an application of and commentary on such security controls without providing a detailed explanation of their workings; it thus assumes that the reader possesses at least a general familiarity with security controls. Readers who are not yet acquainted with such controls may wish to consult a comprehensive catalog such as that found in *NIST Special Publication 800-53, Revision 4,* or *ISO/IEC 27001:2013.*[1]

Approaches to categorizing security controls

Some researchers categorize controls as either **administrative** (i.e., comprising organizational policies and procedures), **physical** (e.g., created by physical barriers, security guards, or the physical isolation of a computer from any network connections), or **logical** (i.e., enforced through software or other computerized decision-making).[2] Other sources have historically classified controls as either **management, operational,** or **technical** controls. In this volume, we follow the lead of texts such as *NIST SP 800-53,* which has removed from its security control catalog the explicit categorization of such measures as management, operational, or technical controls, due to the fact that many controls incorporate aspects of more than one category, and it would be arbitrary to identify them with just a single category.[3] Here we instead utilize a classification of such measures as **preventive, detective,** or **corrective and compensating** controls. This chapter considers the first type of control, while the latter two types are investigated in the subsequent chapters.

Role of security controls in the system development life cycle

The preventive controls discussed in the following sections are organized according to the stage within the process of developing and deploying neuroprosthetic technologies when attention to a particular control becomes most relevant. These phases are reflected in a system development life cycle (SDLC) whose five stages are (1) supersystem planning; (2) device design and manufacture; (3) device deployment in the host-device system and broader supersystem; (4) device operation within the host-device system and supersystem; and (5) device disconnection, removal, and disposal.[4] Many controls relate to more than one stage of the process: for example, the decision to develop a particular control and the formulation of its basic purpose may be developed in one stage, while the details of the control are designed in a later

---

[1] See *NIST Special Publication 800-53, Revision 4: Security and Privacy Controls for Federal Information Systems and Organizations* (2013) and *ISO/IEC 27001:2013, Information technology – Security techniques – Information security management systems – Requirements* (2013).

[2] Rao & Nayak, *The InfoSec Handbook* (2014), pp. 66-69.

[3] See *NIST SP 800-53* (2013).

[4] A four-stage SDLC for health care information systems is described in Wager et al., *Health Care Information Systems: A Practical Approach for Health Care Management* (2013), a four-stage SDLC for an open eHealth ecosystem in Benedict & Schlieter, "Governance Guidelines for Digital Healthcare Ecosystems" (2015), pp. 236-37, and a generalized five-stage SDLC for information systems in *NIST Special Publication 800-100: Information Security Handbook: A Guide for Managers* (2006), pp. 19-25. These are synthesized and applied to create a five-stage SDLC for information systems incorporating brain-computer interfaces in Gladden, "Managing the Ethical Dimensions of Brain-Computer Interfaces in eHealth: An SDLC-based Approach" (2016).

stage and the control's mechanisms are implemented in yet another stage. Here we have attempted to locate a control in the SDLC stage in which decisions or actions are undertaken that have the greatest impact on the success or failure of the given control. This stage-by-stage discussion of preventive controls begins below.

SDLC stage 1: supersystem planning

The first stage in the system development life cycle involves high-level planning of an implantable neuroprosthetic device's basic capacities and functional role, its relationship to its human host (with whom it creates a biocybernetic host-device system), and its role within the larger 'supersystem' that comprises the organizational setting and broader environment within which the device and its host operate. The development of security controls in this stage of the SDLC typically involves a neuroprosthetically augmented information system's designer, manufacturer, and eventual institutional operator. Such controls are considered below.

A. Security planning

1. Centralized management of security planning

In the case of neuroprostheses operated by organizations with relevant technical and managerial capacity, it is feasible to maintain a single coherent, organization-wide system for designing, implementing, and managing security controls and processes.[5] In the case of neuroprosthetic devices that are sold to the general public as consumer electronics devices, the host-device system constituted by an implanted device and its human host may not possess a single coherent, centrally-organized InfoSec approach but may instead reflect a patchwork of diverse and unrelated (and potentially contradictory) information security mechanisms and procedures developed by the device's manufacturer, its OS developer, the application developers of programs installed on the device, and the device's human host.[6]

2. Budgeting and allocation of financial resources

When considering information security for neuroprosthetic devices, special care must be given in the case of devices that will be permanently implanted and may reside within and interact with the biological systems and

---

[5] *NIST SP 800-53* (2013), p. F–144.
[6] See the device ontology in Chapter One of Gladden, *Neuroprosthetic Supersystems Architecture* (2017), and Gladden, "Managing the Ethical Dimensions of Brain-Computer Interfaces in eHealth" (2016), for a list of such relevant parties that can impact a neuroprosthetic device's information security, and see Chapter Five of this book for a discussion of the roles and responsibilities of such parties.

cognitive processes of their human host for a period of years or decades – in order to ensure that a device's operator and (in a case in which the operator might declare bankruptcy, otherwise become incapable of providing InfoSec services, or otherwise fail in its obligation to ensure the long-term safety and information security of the implanted device) the device's human host will be able to provide the resources needed to ensure the safe and secure long-term functioning of the device.[7]

3. Planning of the system development life cycle

The system development life cycle[8] for neuroprosthetic devices should take into account the fact that once a device has been implanted in a human host, the organization operating the device may lose control over some or all aspects of the operations and maintenance phase and, in particular, the disposal phase for the device – insofar as it may not be legally, ethically, or practically feasible to carry out some kinds of activities (such as physically altering a device's integration with the neural circuitry of its human host or subjecting a device to removal or recall) without the host's consent.

4. Development of a system security plan

Development of an effective system security plan[9] for certain kinds of neuroprosthetic devices may be complicated by the fact that the human host in whom a device is implanted is either not aware of the device's existence or is not able – due to legal, ethical, or practical considerations – to participate constructively in execution of the system security plan. This may be the case, for example, in situations in which devices have been implanted in children, persons who are in a coma, or individuals who are otherwise incapacitated, as well as in the case of some devices used for military or intelligence-gathering purposes in which a host's access to information about a device and its operations must be constrained. Significant questions relating to personal privacy, human autonomy, bioethics, and the ethics of technology arise in such situations that must be addressed.[10]

5. Formulation of an information security architecture to support the enterprise architecture

An information security architecture is designed to describe in a clear and coherent manner "the overall philosophy, requirements, and approach to be

---

[7] Regarding the allocation of resources, see *NIST SP 800-53* (2013), p. F–156.

[8] See *NIST SP 800-53* (2013), p. F–157 for additional discussion of an SDLC in the context of information security.

[9] *NIST SP 800-53* (2013), pp. F–139-41.

[10] See Bowman et al., "The Societal Reality of That Which Was Once Science Fiction" (2012), for a discussion of some such issues.

taken with regard to protecting the confidentiality, integrity, and availability of organizational information" and to explain "how the information security architecture is integrated into and supports the enterprise architecture."[11] For an advanced neuroprosthetic device, the information security architecture not only supports and is integrated into the overall enterprise architecture of the organization operating the device but must also incorporate (or, in effect, function as) the biomedical security architecture and cognitive and noetic security architecture of the human being in whom the device is implanted.

6. Formulation of a security concept of operations (CONOPS)

An organization's security concept of operations (or CONOPS) for an information system typically describes "how the organization intends to operate the system from the perspective of information security;"[12] any changes to the ongoing operations relating to the information system (and thus its CONOPS) will eventually be reflected in an updated system security plan, information security architecture, or other documents such as information security specifications governing specifications for future hardware and software acquisitions, SDLC materials, and systems engineering materials.[13] In the case of advanced neuroprostheses, the CONOPS may also draw on (and changes to the CONOPS may need to be reflected in):

- **Biomedical and bioengineering specifications** that set operating parameters that should be maintained within the biological organism of a device's human host in order to ensure its safe and effective functioning.

- **Biocybernetic system architecture** documents that describe the processes of communication and control within and between the device and its human host.

- **Cognitive and noetic security architecture** plans which describe and dictate the ways in which the privacy and autonomy of the mind of the device's human host (and the mind's integral cognitive processes) will be ensured.

7. Formalization of operations security (OPSEC) practices and personnel

Operations security (or OPSEC) attempts to secure an organization's sensitive information not directly through the development of access controls for the information itself but by ensuring more generally that the organization's operations do not unnecessarily disclose information that could be

---

[11] *NIST SP 800-53* (2013), p. F–142.

[12] *NIST SP 800-53* (2013), p. F–142.

[13] *NIST SP 800-53* (2013), p. F–142.

used by adversaries to develop more effective means of attacking the organization and acquiring the sensitive information that they ultimately wish to obtain. OPSEC carries out its work through the "(i) identification of critical information (e.g., the security categorization process); (ii) analysis of threats; (iii) analysis of vulnerabilities; (iv) assessment of risks; and (v) the application of appropriate countermeasures."[14]

In the case of advanced neuroprosthetic devices, the mandate of OPSEC practices and personnel may also need to be broadened to include not only preventing the unnecessary disclosure of information about an organization's internal operations but also preventing the unnecessary disclosure of information about the personal (non-organizational) activities of members of the organization who possess neuroprostheses, insofar as the sensitive organizational information contained in such devices could potentially be targeted through attacks that utilize avenues relating to members' private lives and activities.[15] At the same time, traditional OPSEC objectives of limiting the dissemination of information about the existence, purpose, use, and context of information systems may sometimes conflict with the desires of neuroprosthetic devices' human hosts, whom it may be legally and ethically difficult to prevent from disclosing information about their personal life and activities, should they desire to do so.

B. Risk assessment and formulation of security requirements

1. Criticality analysis of devices and components

In the case of an advanced neuroprosthetic devices, some device components might be designated as critical[16] not because they directly secure information contained within a device but because they support the physical and psychological health and safety of the device's human host, thereby indirectly ensuring the security of information held within the natural cognitive processes of the host's mind.[17]

---

[14] See *NIST SP 800-53* (2013), p. F–210, for a general description of OPSEC practices and personnel.

[15] See Chapter Three of this volume for a discussion of the ways in which a neuroprosthetic device is inextricably entangled with the larger host-device system in which it operates, through integration into the neural circuitry of its human host.

[16] *NIST SP 800-53* (2013), p. F–174.

[17] See Chapter Three of this text for the need to secure both a device and its larger host-device system. For the way in which a neuroprosthesis might enhance the information security of its host by, e.g., counteracting the effects of degenerative neurological conditions affecting memory and cognition, see Gladden, "Information Security Concerns as a Catalyst for the Development of Implantable Cognitive Neuroprostheses" (2016).

2. Security categorization of the system and its information

For a neuroprosthetically augmented information system, security categorization[18] includes classifying the system and its information not only in relation to requirements determined by laws, regulations, and organizational policies and ethical standards relating to computer equipment but also in accordance with those regulations and guidelines that apply to personal health information, implantable medical devices, surgical procedures, psychological diagnostic and therapeutic procedures, and other relevant fields.[19]

3. Threat modelling and vulnerability analysis

In some cases, the ability of developers to perform effective threat modeling and vulnerability analysis[20] for advanced neuroprosthetic devices (or their constituent components or software) that are under development may be impeded by the fact that a device's functional and operational characteristics do not become clear until it is implanted in a particular human host and integrated with the host's neural circuitry, as a device may be largely passive in nature and its functional characteristics determined largely by the unique traits (e.g., memories, thoughts, or volitions) of the mind of its human host.[21]

4. Risk assessment

A risk assessment should be carried out (and updated as needed) for an information system as a whole as well as for relevant component devices and particular uses of the system, in order to analyze the "risk, including the likelihood and magnitude of harm, from the unauthorized access, use, disclosure, disruption, modification, or destruction of the information system and the information it processes, stores, or transmits."[22] Such assessments should identify both risks resulting from factors internal to the organization that will

---

[18] *NIST SP 800-53* (2013), p. F–151.

[19] For an overview of ethical issues with ICT implants – many of which are relevant for advanced neuroprosthetics – see Hildebrandt & Anrig, "Ethical Implications of ICT Implants" (2012). For ethical issues in information security more generally, see Brey, "Ethical Aspects of Information Security and Privacy" (2007). For regulatory issues, see Kosta & Bowman, "Implanting Implications: Data Protection Challenges Arising from the Use of Human ICT Implants" (2012); McGee, "Bioelectronics and Implanted Devices" (2008); Mak, "Ethical Values for E-Society: Information, Security and Privacy" (2010); McGrath & Scanaill, "Regulations and Standards: Considerations for Sensor Technologies" (2013); and Shoniregun et al., "Introduction to E-Healthcare Information Security" (2010).

[20] *NIST SP 800-53* (2013), p. F–175.

[21] See Chapter One of this text for a discussion of passive neuroprostheses and Chapter Three for a discussion of the classification of vulnerabilities and threats relating to advanced neuroprosthetic devices.

[22] *NIST SP 800-53* (2013), p. F–152.

operate the device and internal to the device itself as well as risks resulting from external factors or agents.

In the case of advanced neuroprostheses, the risk assessment should not be limited to evaluating the potential impact of the unauthorized use or alteration of information contained within a device itself but also the impacts of the possible unauthorized use or alteration of information (such as memories or sense data) that are not contained within the physical components of the device but which are received, generated, transmitted, or stored by natural biological systems that belong to the device's human host and which are thus contained within the host-device system.[23]

5. Formulation of resource availability priorities and guarantees

Priority protection[24] may be utilized, for example, to ensure that a neuroprosthetic device's processes that control and enable the proper functioning of the respiratory and circulatory systems of its human host's body enjoy higher-priority access to the device's resources than processes that provide an augmented-reality enhancement to the user's vision that is useful but not critically necessary. Quotas[25] may be utilized to ensure that a particular process does not consume excessive resources, even when there is no other immediate demand for the resources; the use of such quotas can help ensure that spare resources are available for instantaneous access should they be required by another process (particularly a high-priority one) which needs the resources immediately in order to execute some critical task or prevent harm to a device's host or user.

6. Defining security requirements for the acquisition process

An organization's acquisition process for advanced neuroprostheses should not only define traditional functional, strength, and assurance requirements[26] relating to information security for a device itself but also for its larger host-device system. Goals for the host-device system also include preserving the cognitive and noetic security, privacy, and autonomy of the device's human host.

---

[23] See Chapter Three of this text for a discussion of information security for a device versus information security for its host-device system.

[24] *NIST SP 800-53* (2013), p. F–187.

[25] *NIST SP 800-53* (2013), p. F–187.

[26] Regarding the formulation of security requirements for the acquisition process, see *NIST SP 800-53* (2013), pp. F–158-60.

C. Personnel controls

1. Separation of duties

Separation of duties[27] may be difficult to implement in cases in which a single person is both the operator and host of a neuroprosthetic device, as well as potentially the developer of applications or other content for use by the device. (Indeed, for some kinds of passive neuroprostheses, the brain of a human host may also be providing the 'operating system' for a device.[28])

2. Risk designations for positions

For advanced neuroprosthetic systems, the risk designation for positions[29] must take into account not only the extent to which a position's occupant will be able to directly access information stored within a device and within the natural biological systems of the device's human host but also the extent to which the position's occupant can indirectly alter, damage, or destroy information stored within the device or its host's biological systems by operating the device or interacting with its host in such a way that affects natural or artificial systems within the host's body that are not directly connected to the neuroprosthetic device but which can have an impact on the confidentiality, availability, or integrity of information stored within the device or its host. For example, the position of a technician who can alter a device's settings in such a way that causes the device's host to enter a comatose state may require a high risk designation, even if the position's occupant does not have any direct ability to retrieve or interpret information stored within the device or the host's natural memory systems.

3. Rules of behavior for organizational personnel

Under normal circumstances, the organization operating an information system may be able to unilaterally update the rules of behavior governing the use of that system and require all members of the organization to produce "a signed acknowledgment from such individuals, indicating that they have read, understand, and agree to abide by the rules of behavior;"[30] an individual who declines to agree to the new rules of behavior may be denied access to the information system, removed from the organization, or potentially subjected to other disciplinary or personnel action as allowable by relevant law, regulations, employment agreements, and ethical guidelines. However, in the case of neuroprosthetic devices that have been implanted in the members of

---

[27] *NIST SP 800-53* (2013), p. F–18.

[28] See Chapter One of this text for a discussion of passive neuroprosthetic devices.

[29] *NIST SP 800-53* (2013), p. F–145.

[30] *NIST SP 800-53* (2013), p. F–141.

an organization and are operated by that organization, it may be illegal, un-ethical, and impractical to attempt to force members to agree to new rules of behavior that are unilaterally imposed by the organization after devices have already been implanted; it may also be impermissible to attempt to deactivate or remove such devices simply because their hosts decline to agree to the updated rules of behavior.

4. Determining dual authorization for the execution of instructions

Controls that require the approval of two different authorized parties be-fore instructions will be executed[31] may be impractical and inappropriate in the case of neuroprosthetic devices that are operated by their human host and which must be able to function when the host is in an open environment where a device cannot be accessed by other parties (e.g., a remote area with-out cell phone service or Internet access).

5. Personnel screening for access to confidential information

In the case of neuroprosthetically augmented information systems, it may sometimes occur that neuroprosthetic devices are used by their operating or-ganization to gather classified or sensitive information that the device's hu-man hosts are not themselves authorized to access or possess;[32] the legal and ethical conditions governing such activities should be carefully clarified.

6. Formalization of access agreements

Organizational access agreements may include components such as "non-disclosure agreements, acceptable use agreements, rules of behavior, and conflict-of-interest agreements."[33] In the case of advanced neuroprosthetic devices, such agreements may be designed to protect the interests of multiple parties; for example, device designers and manufacturers, OS and software developers, and device operators may wish to ensure that devices' human hosts will not misuse information that they acquire through their possession of and interaction with the devices, and the devices' hosts may wish to ensure the confidentiality, possession, and legal ownership of sensitive information (e.g., relating to their biological or cognitive processes) that the devices may acquire or information (such as ideas, memories, inventions, discoveries, or artistic creations) that may be generated by the host's mind with the assis-tance of a device.

---

[31] *NIST SP 800-53* (2013), p. F–11.

[32] Regarding personnel policies governing access to confidential information, see *NIST SP 800-53* (2013), p. F–146.

[33] *NIST SP 800-53* (2013), p. F–148.

7. Formal indoctrination of personnel

Formal training for the hosts of neuroprosthetic devices regarding the legal and ethical frameworks governing the functioning of their devices and their relationships to classified or sensitive information may be necessary,[34] for example, if a host possesses an artificial eye or other neuroprosthesis that is continuously recording, uploading, and potentially making publically available information received from external environmental phenomena surrounding the host or from the host's internal cognitive processes.

8. Training against insider threats

In the case of neuroprosthetic devices that allow direct access to the cognitive processes (including sensory perceptions, thoughts, or memories) of their human host, complex legal and ethical questions arise over the proprietary of the accessing of such information by an organization's personnel in order to assess whether the host may constitute an insider threat to the organization's information security.[35]

9. Establishment of probationary periods

The use of probationary periods[36] for individuals receiving authorized access to information systems may not be legally, ethically, or practically feasible in the case of hosts in whom neuroprosthetic devices are being implanted. Any 'probationary period' designed to ensure a host's knowledge of and commitment to organizational policies (including InfoSec practices) may need to take place before the device is implanted and integrated into the host's neural circuitry, as it could be impossible to remove or deactivate the device after its implantation if the host should not successfully complete the probationary period. On the other hand, any probationary period designed to test a host's ability to successfully operate a device and use it to perform necessary InfoSec-related tasks may necessarily need to take place after the device has been implanted (and after the host has undergone any required recovery, adaptation, and training period), as it may be impossible to simulate operational conditions or to fully train and test the host in the device's use prior to the device's physical integration with the host's neural circuitry.

---

[34] Regarding InfoSec policies relating to the formal indoctrination of personnel, see *NIST SP 800-53* (2013), p. F–146.

[35] Regarding insider threats, see *NIST SP 800-53*, Rev. 4 (2013), p. F–38, and McCormick, "Data Theft: A Prototypical Insider Threat" (2008). See Chapter Two of this text for a discussion of other insiders within an organization who might pose a threat to a neuroprosthetic device's host or operator.

[36] *NIST SP 800-53* (2013), p. F–223.

10. Establishing personnel sanctions for the violation of InfoSec policies

On the one hand, in order to ensure the security of highly sensitive information regarding the biological processes or cognitive activity of human hosts in whom neuroprosthetic devices are implanted, it may be necessary to enact a stringent formal sanctions[37] process to discipline individuals within an organization who disregard or violate established InfoSec procedures. On the other hand, there may be significant legal and ethical issues that complicate an organization's ability to discipline the human host of a neuroprosthetic device who violates organizational information security procedures, especially if those procedural requirements have been unilaterally imposed by an organization subsequent to a device's implantation or otherwise enacted without the full informed consent of the device's host.

11. Establishing personnel termination procedures

Standard procedures upon termination of the employment of an organization's member may include action by the organization that "Disables information system access" previously enjoyed by the employee within a specified time period, "Terminates/revokes any authenticators/credentials associated with the individual," "Retrieves all security-related organizational information system-related property," and "Retains access to organizational information and information systems formerly controlled by terminated individual."[38] The ability of an organization to carry out such actions in the case of an organizational information system that takes the form of a neuroprosthetic device implanted in a (former) employee may be significantly constrained by legal, ethical, and practical considerations.

For example, even if an employee had signed an employment contract or agreement clearly specifying that any devices subsequently implanted in the employee by the organization (and all information contained within them) were property of the organization and that the organization enjoyed the right to reclaim such devices at information any time, the ability of the organization to enforce the agreement and reclaim an implanted device through forcible surgical extraction would be legally and ethically doubtful, at best – although the employee could potentially be subject to civil action for theft or conversion. Moreover, even if it is technologically possible for an organization to execute such actions by sending remote instructions to a device, the organization may not have a legal or ethical ability to forcibly reclaim a neuroprosthetic device or to destroy all of the information contained on it (such

---

[37] *NIST SP 800-53* (2013), p. F–150.
[38] *NIST SP 800-53* (2013), p. F–147.

as thoughts or memories of the device's host), even if an employment agreement were unilaterally breached or terminated by the device's host in contravention of the agreement's terms and conditions.

12. Establishing post-employment obligations of personnel

Automated systems may be employed to ensure that former employees are, for example, not able to make use of classified or sensitive information contained within implanted neuroprostheses that had been provided to them by their former employer for work-related purposes, or even to utilize the devices at all.[39] If a neuroprosthetic device has been designed and constructed in such a way that ongoing proactive authorization or support from the organization employing the device's host (e.g., wireless signals sent to the device from an external organizational information system) are required in order for the device to function or for its contained information to be accessible, the withdrawal of such authorization upon termination of an employee may constitute a practice that is legally and ethically permissible, provided that it does not have a negative impact on the employee's psychological, physical, or social wellobeing. In its natural state (i.e., in the absence of such authorization signals) the device will simply become nonfunctional, and the organization has no obligation to provide such authorizations.[40] On the other hand, the situation becomes more legally and ethically complex if an implanted neuroprosthetic device's natural state is one in which the device functions nominally and information contained within it is available to the device's host, and the functioning of the device (and availability of its contained information) can only be suppressed by the organization through the ongoing application of some proactive measure – such as bombarding the host's body with electromagnetic impulses that jam the device's communications or otherwise disrupt its operation. The legal, ethical, and practical ability of an organization to carry out such measures to impair the operation of an implanted neuroprosthesis may be severely constrained.

---

[39] Regarding post-employment requirements relating to information security, see *NIST SP 800-53* (2013), p. F–147.

[40] See the discussion in Chapter Three of this text of proposed schemes that utilize external hardware tokens, cloaking devices, or gateway devices whose presence causes an implanted medical device to behave in a particular way during emergency (or non-emergency) situations. Employers could potentially develop similar systems in which an employer could disable an implanted neuroprosthesis or cause it to 'fail closed' not by physically accessing the device or even wirelessly sending the device a command to disable itself (which may not be legally or ethically possible) but simply by confiscating, disabling, or failing to renew some external token or device in the possession of the implanted neuroprosthetic device's host that is needed in order to prolong the functioning of the implanted device.

D. Designing an architecture for the entire information system

1. Development of comprehensive information system documentation

An organization must acquire and appropriately secure documentation from a neuroprosthetic device's designers, manufacturers, and OS developers regarding subjects such as high-level design principles, low-level design details, the functional properties of security controls built into the device and its OS, source code, external system interfaces, and the full characteristics of administrative accounts built into the device and its OS.[41] In the case of an advanced neuroprosthesis, such documentation would also include recommendations and requirements regarding the biological systems and structures into which the device will be integrated, recommended methods for performing implantation and integration of the device into the host's neural circuitry, and circumstances in which implantation of a device into a particular host is contraindicated or deactivation or removal of the device would be required.

2. Intentional heterogeneity of systems, devices, and components

Increasing the heterogeneity and diversity of the sources, forms, functionalities, and procedures relating to the individual components of neuroprosthetic devices or the larger information systems that incorporate them is a double-edged sword: on the one hand, such diversity "reduces the likelihood that the means adversaries use to compromise one information system component will be equally effective against other system components, thus further increasing the adversary work factor to successfully complete planned cyber attacks."[42] On the other hand, increased heterogeneity and diversity of systems and components may contravene the InfoSec principle of utilizing conceptually simple design and may increase the cost, complexity, and difficulty of properly managing information systems.[43]

3. Planning of connections to non-organizationally owned systems and devices

Neuroprosthetic devices may operate within a complex context in which, for example, key components of a device are owned by an organization but other components consist of biological matter that is a part of a host's body. This may complicate the process of establishing controls for the sharing of information with components or systems that are not owned by the organization.[44]

---

[41] *NIST SP 800-53* (2013), pp. F–160-61.

[42] *NIST SP 800-53* (2013), p. F–204.

[43] *NIST SP 800-53* (2013), p. F–204.

[44] Regarding non-organizationally owned systems, devices, and components, see *NIST SP 800-53* (2013), p. F–33.

4. Analysis of the reliance on external information systems

Insofar as the information within a neuroprosthetically augmented information system can be accessed by the mind of a device's host, it is possible (and in some situations perhaps likely) that the information will eventually be transmitted to or copied – in a manner that may or may not accurately represent the original source information – into external information systems to which the host's mind has access (such as the host's personal computer, smartphone, or other systems).[45]

5. Planning of boundary protection for physical, psychological, and logical boundaries

Boundary protection is of critical importance for advanced neuroprosthetic devices and takes on new meanings in this context. An information system should monitor (and, as appropriate, control) all communications taking place at the system's external boundary which cause data to enter or leave the system as well as communications taking place across key internal boundaries within a system or its constituent components.[46] In the case of advanced neuroprosthetic devices, key boundaries include the:

- Physical boundary between a neuroprosthetic device and the biological matter of its human host (and in particular, the physical boundary or interface between the device and natural biological neurons within the host's body).[47]

- Physical boundary between a neuroprosthetic device and the environment external to its host's body (for prostheses that are exposed to the external environment). This may include a boundary with particular systems, devices, or individuals located within that external environment.

- Physical boundary between a host's body and the surrounding external environment.[48]

- Physical boundary between a neuroprosthetic device and other implanted devices within its host's body.

---

[45] Regarding the use of external information systems, see *NIST SP 800-53* (2013), p. F–32.

[46] *NIST SP 800-53* (2013), p. F–188.

[47] See Chapter One of this text and the device ontology in Chapter One of Gladden (2017) for a discussion of different kinds of neural interfaces.

[48] For a discussion of the significance of the physical boundaries of a human organism and the ways in which technologies such as implantable neuroprostheses can impact cognitive processes and the "moral sense of person" versus "the notion of person as a subject of experiences," see Buller, "Neurotechnology, Invasiveness and the Extended Mind" (2011).

- Physical, physiological, psychological, and logical boundaries between different neuronal processes within a host-device system – such as the boundary at which environmental stimuli are transduced into electrochemical signals by sensory organs and the boundary at which raw sense data is transformed into sensory perceptions within the host's mind. Also included are psychological boundaries between phenomena such as memory, emotion, volition, and conscience whose phenomenological[49] and experiential boundaries as seen from the perspective of the mind of a device's human host may not clearly correspond to physical boundaries within the host-device system.

Some neuroprosthetic devices may interact with their human host only through a small number of managed interfaces (e.g., synaptic connections through which signals are received and transmitted); other devices, such as those composed of biological material, may interact with a device's host through unmanaged interfaces that may change significantly over time as a result of the growth of biological components of the device, changes in the host's organism, or both. It may be appropriate and desirable to create outward-facing subnetworks (or '**demilitarized zones**') that are separated physically or logically from a device's internal networks[50] and which interface either with the biological systems and cognitive processes of the device's human host, with supplemental prostheses or other accessories that can be connected to the device (e.g., through generic ports or sockets), with the external physical environment (e.g., to prevent sensory overload or 'sense hacking' that could occur if the external environment supplied stimuli directly to internal systems), or with other external systems such as Wi-Fi networks and the Internet.

6. Planning of internal system interconnections

If multiple kinds of neuroprostheses are available for acquisition, installation, and use by members of the general public as consumer electronics devices, it may be difficult or impossible to predict the ways in which multiple devices may be combined and interconnected within a single human host. Even if devices do not directly interconnect with one another, they may indirectly interconnect through the host's brain and mind, which can serve as a bridge allowing information to flow between devices and influence one another.[51]

---

[49] See Chapter Two of Gladden (2017) for an analysis of such boundaries from a biocybernetic perspective. For an exploration of phenomenological issues, see Heersmink, "Embodied Tools, Cognitive Tools and Brain-Computer Interfaces" (2011).

[50] *NIST SP 800-53* (2013), p. F–188.

[51] Regarding system interconnections, see *NIST SP 800-53* (2013), pp. F–57-58.

7. Planning the physical partitioning of the information system

Typical approaches to information system partitioning[52] – such as physically separating different components in different racks within the same room, in different rooms, or in wholly different geographical locations – may not be feasible in the case of an implantable neuroprosthetic device that must be as small and compact as possible. It may be possible to physically separate components through the creation of body area networks (BANs) or body sensory networks (BSNs)[53] whose components are distributed throughout a host's body and interact with one another wirelessly. It may also be possible to separate the implantable device from external devices or support systems that communicate with the implantable device; however, care must be taken to ensure that the implanted portion of the system can continue to operate in a way that will not create the danger of physical or psychological harm for the device's host or others if the device should temporarily or permanently lose the ability to communicate with the external systems (e.g., because the device's host has entered a building whose construction blocks the transmission of wireless signals).

8. Planning of hardware separation

Hardware separation mechanisms[54] may be utilized, for example, to segregate the systems of a neuroprosthetic device that interact directly with the neural circuitry of the device's host from those which relate to the device's power supply or wireless communication with external support systems.

9. Planning of application partitioning

Separating a system's user functionality and interface services from its administrative and system management functionality[55] may be difficult or impossible in the case of a neuroprosthetic device whose human host is also its operator. In other cases, such partitioning may be not only desirable but necessary – for example, if a device's host is the 'user' responsible for controlling some aspects of the device's ongoing operation but management of key med-

---

[52] *NIST SP 800-53* (2013), p. F–207.

[53] See Ullah et al., "A Study of Implanted and Wearable Body Sensor Network" (2008); Cho & Lee, "Biometric Based Secure Communications without Pre-Deployed Key for Biosensor Implanted in Body Sensor Networks" (2012); and Li et al., "Advances and Challenges in Body Area Network" (2011).

[54] *NIST SP 800-53* (2013), p. F–185.

[55] *NIST SP 800-53* (2013), p. F–184.

ical and technical aspects of the device's behavior is controlled by remote 'users' in the form of a team of specialized medical and IT personnel who possess expert knowledge that the human host lacks.[56]

10. Determining an architecture for device name and address resolution

Particularly in the case of a neuroprosthetic system that includes multiple devices implanted within a single human host that must communicate with one another,[57] it may be appropriate and desirable to utilize separate name and address resolution services[58] (such as those offered by DNS servers and network routers) for processing internal information requests from the component devices that constitute the system and external information requests from external networks such as the Internet.

11. Designing host-client device systems

In the sense commonly employed within the field of IT management, the word 'host' does not refer to the human being in whom a neuroprosthetic device is physically implanted but to a device (such as a server or desktop computer) that executes some piece of software and potentially serves as a host in a host-client device system.[59] In the case of an organization that has deployed many neuroprosthetic devices among its personnel, the organization may operate a centralized information system housed within a secure organizational facility that serves as the host for the client neuroprosthetic devices. If all of the client devices are monitored or controlled by, receive software updates from, or are otherwise affected by the centralized system, the use of effective security controls to protect that core system is important for securing its client devices.

12. Planning of collaborative computing capacities

Advanced neuroprosthetic devices may be able to serve directly as collaborative computing devices;[60] for example, an artificial eye implanted in one human host could potentially stream live video that can be viewed by other persons in order to share in the host's visual experiences. Similarly, a human host possessing a body with robotic cybernetic limbs might allow a professional dancer to take temporary control of the body in order to create a form of shared performance art. Possibilities also exist for the internal cognitive

---

[56] See Chapter Three of this text for a discussion of the distinction between a neuroprosthetic device's human host and its user or users.

[57] Body area networks and body sensor networks typically constitute such systems. See Ullah et al. (2008); Cho & Lee (2012); and Li et al. (2011).

[58] *NIST SP 800-53* (2013), p. F–201.

[59] *NIST SP 800-53* (2013), p. F–223.

[60] *NIST SP 800-53* (2013), p. F–197.

processes of neuroprosthetic devices' hosts to form collaborative computing devices by creating 'hive minds' or communities of individuals whose minds are linked through their implanted neuroprostheses.[61] A neuroprosthetic device may also allow its human host new ways of accessing, controlling, and obtaining information from traditional collaborative computing devices such as cameras, microphones, or printers. The growth of the Internet of Things and new kinds of devices such as 3D printers and smart homes creates entirely new types of networked systems that can potentially be accessed and controlled by means of neuroprosthetic devices.[62] All of these possibilities raise significant questions of information security both for the users and operators of neuroprosthetic devices and for other individuals who use the collaborative computing devices or can be affected by their activities.

13. Planning for information in shared resources

Care must be given to ensuring information security in situations in which other users, accounts, or processes may have access to shared system resources through which information created or used by or otherwise related to a neuroprosthetic device has passed or within which it has been stored. In such circumstances, information security is pursued through the control of **object reuse** and **residual information protection**.[63] Similar but distinct concerns include the need to address situations of **information remanence** in which action has been undertaken to erase or destroy information (and it may nominally be designated by a system as 'deleted') but residual traces of the data still exist and can potentially be accessed,[64] as well as situations in which **covert channels** are utilized to access, transmit, or manipulate information in ways

---

[61] The prospect of creating 'hive minds' and neuroprosthetically facilitated collective intelligences is investigated, e.g., in McIntosh, "The Transhuman Security Dilemma" (2010); Roden, *Posthuman Life: Philosophy at the Edge of the Human* (2014), p. 39; and Gladden, "Utopias and Dystopias as Cybernetic Information Systems: Envisioning the Posthuman Neuropolity" (2015). For critical perspectives on the notion of hive minds, see, e.g., Maguire & McGee, "Implantable brain chips? Time for debate" (1999); Bendle, "Teleportation, cyborgs and the posthuman ideology" (2002); and Heylighen, "The Global Brain as a New Utopia" (2002).

[62] See Evans, "The Internet of Everything: How More Relevant and Valuable Connections Will Change the World" (2012); Merkel et al., "Central Neural Prostheses" (2007); and Gladden, "Neural Implants as Gateways to Digital-Physical Ecosystems and Posthuman Socioeconomic Interaction" (2016).

[63] *NIST SP 800-53* (2013), p. F–186.

[64] With regard to the case of information stored within a physical neural network – and perhaps even within the human brain's natural biological long-term memory storage systems – researchers have had some success with attempting to manipulate or delete specific memories stored within the brains of mice; see, e.g., Han et al., "Selective Erasure of a Fear Memory" (2009). However, it is unclear to what extent, if any, it might someday be possible to 'delete' or 'erase' complex

that bypass information flow restrictions – potentially by employing a device's systems, components, or processes in imaginative or counterintuitive ways that were never anticipated by a device's designer.[65]

For some neuroprosthetic devices, 'shared resources' may include biological systems of a device's human host (such as the circulatory system, sensory organs, or limbs) or the host's cognitive systems and processes (such as natural memory storage systems and particular mnemonic content); the use of a neuroprosthetic device may create traces of information in such shared systems that can be accessed by other processes or users of the device or other implanted devices within the host's body, even if they lack direct access to components, user accounts, or processes within the neuroprosthetic device that created the original information.

14. Planning of offline storage

For some kinds of implantable neuroprosthetic devices (e.g., those that store information within an internal physical neural network or which lack mechanisms for transmitting information to or receiving information from external systems), a device itself and its own internal storage mechanisms may constitute a form of off-line storage, insofar as the information is not accessible from any sort of external networks.[66]

15. Planning of out-of-band channels

For human beings possessing certain kinds of neuroprosthetic devices, such a device may provide a new kind of 'out-of-band channel'[67] for conveying information directly to the conscious awareness or cognitive processes of its human host in a way that bypasses or avoids the traditional biologically based 'in-band channels' comprising sensory organs. Conversely, for the human host of a sensory or cognitive neuroprosthesis who ordinarily receives sensitive or secure information through the device (e.g., with information being presented in the host's visual field through use of augmented reality or being directly incorporated into the host's short- or long-term memory), receiving information through the use of the host's natural biological sensory organs may constitute the use of an out-of-band channel.

---

long-term memories stored within the natural long-term memory systems of human brains; information remanence may thus become a major challenge for neuroprosthetic devices utilizing physical neural networks, biological components, and engrams for the storage of information. Such issues surrounding the possibility of deleting long-term memories may become even more vexing if, e.g., holographic or holonomic models of the brain's memory systems are correct.

[65] *NIST SP 800-53* (2013), p. F–186.

[66] Regarding offline storage, see *NIST SP 800-53* (2013), p. F–204.

[67] For the InfoSec implications of out-of-band channels, see *NIST SP 800-53* (2013), pp. F–209-10.

SDLC stage 2: device design and manufacture

The second stage in the system development life cycle includes the design and manufacture of a neuroprosthetic device and other hardware and software that form part of any larger information system to which the device belongs. The development of security controls in this stage of the SDLC is typically performed by a device's designer and manufacturer, potentially with instructions or other input from the system's eventual operator. Such controls are considered below.

A. General device design principles

1. Formal InfoSec policy modelling for design of a device, system, and supersystem

The kinds of formal policy modelling tools traditionally used to model practices such as nondiscretionary access control policies with formal languages[68] may have limited applicability for some kinds of advanced neuroprostheses. For example, some kinds of neuroprosthetic devices that comprise physical neural networks or swarms of nanorobotic elements may not include nondiscretionary access controls that can easily be modelled; in the case of devices that are passively controlled by the minds and cognitive processes of their human hosts, a system's security controls may be entirely discretionary and controlled by the decision-making and volition of the device's human host. It may be possible to develop new kinds of formal policy models and modelling languages that address the unique information security situations of advanced neuroprosthetic devices (including the typically important goal of preserving autonomy and agency for the host-device system as a whole).

2. Updating of security engineering principles

In developing the designs and specifications for advanced neuroprosthetic devices, entirely new kinds of information system security engineering principles[69] may need to be developed that incorporate considerations relating to cognitive and noetic security and the preservation of human agency and autonomy within a host-device system.

3. Pursuit of trustworthiness through security functionality and assurance

The trustworthiness[70] of an information system depends both on the (1) set of features, mechanisms, and procedures built into constituent devices

---

[68] *NIST SP 800-53* (2013), p. F–178.
[69] *NIST SP 800-53* (2013), p. F–162.
[70] *NIST SP 800-53* (2013), p. F–173.

and the operating environment that together constitute the system's *security functionality* and (2) the *security assurance* that allows an organization to believe that the potential benefits offered by the security functionality are actually being obtained through a proper and effective implementation of the functionality.[71]

It should be noted that in at least some respects, some kinds of advanced neuroprostheses may be inherently *untrustworthy*. For example, certain kinds of devices that include a physical neural network and which interact closely with the natural memory systems of the human mind to expand or support the mind's long-term memory storage may be subject to the same kind of mnemonic compression, distortion, and gradual information degradation that is observed with natural human memories.[72]

4. Use of conceptually simple design

Requirements that developers develop systems that utilize "a complete, conceptually simple protection mechanism with precisely defined semantics"[73] may be difficult to realize in situations in which protection mechanisms may, for example, be implemented and directed largely in a discretionary manner by the mind of the human host in whom a device is implanted.

5. Design of coupled and cohesive security function modules

It is a best practice to develop and utilize "security functions as largely independent modules that maximize internal cohesiveness within modules and minimize coupling between modules."[74] In the case of highly sophisticated multimodal neuroprosthetic devices, it may be possible to develop individual security functions that separately address, for example, security relating to incoming sense data (with data from each sensory organ handled separately), internal cognitive activities (with each activity possessing its own security functions), and outgoing motor instructions (with different security functions for each motor modality and effector. The development of independent modules may not be possible with other kinds of neuroprosthetic devices, such as those that utilize a physical neural network or which store and process information holographically.

---

[71] *NIST SP 800-53* (2013), p. F–173.

[72] For a discussion of such issues, see Dudai, "The Neurobiology of Consolidations, Or, How Stable Is the Engram?" (2004).

[73] *NIST SP 800-53* (2013), p. F–179.

[74] *NIST SP 800-53* (2013), p. F–186.

6. Planning non-persistence and the regular refreshing of devices and components

Rather than waiting until it has been detected that particular components or services have been compromised and then replacing them, terminating their functionality, or otherwise addressing the situation, an organization may proactively refresh components and services at regular or random intervals. Such procedures can reduce the effectiveness of certain kinds of **advanced persistent threats** (APTs) that must have access to or operate within a particular computing environment for a substantial period of time in order to successfully exploit vulnerabilities and complete their attack.[75] For some kinds of neuroprosthetic devices, non-persistence may be difficult to implement, insofar as a device must provide continual service and 100% availability in order to avoid causing physical or psychological harm for its host or operator, and the time and actions needed to refresh components or services would cause an impermissible interruption or disruption to the device's functionality.[76] In other cases, it may be possible to refresh components or services during non-critical moments (e.g., when a device's host is asleep or not engaging in particular kinds of activities). Other kinds of neuroprosthetic devices (such as those utilizing biological components or neural networks) may neither require nor allow such periodic refreshing of components or services.

7. Planning physical and logical separation of information flows

For some kinds of neuroprostheses (e.g., those utilizing physical neural networks) it may be extremely difficult to segregate different kinds of information moving through the devices.[77]

8. Denial of inbound and outbound communications by default

The practice of denying all inbound and outbound network communications traffic by default and allowing it only after it has been approved as an

---

[75] *NIST SP 800-53* (2013), p. F–232.

[76] See Chapter Three of this text for a discussion of neuroprosthetic devices for which 100% availability is required and any downtime presents a major hazard.

[77] For example, if various holographic or holonomic models of the human brain's cognitive processing and memory storage are correct, it may be difficult or impossible to isolate a certain small group of neurons as completely 'containing' a particular memory or thought. For discussion of such issues, see, e.g., Longuet-Higgins, "Holographic Model of Temporal Recall" (1968); Westlake, "The possibilities of neural holographic processes within the brain" (1970); Pribram, "Prolegomenon for a Holonomic Brain Theory" (1990); and Pribram & Meade, "Conscious Awareness: Processing in the Synaptodendritic Web – The Correlation of Neuron Density with Brain Size" (1999). An overview of conventional contemporary models of long-term memory is found in Rutherford et al., "Long-Term Memory: Encoding to Retrieval" (2012). Regarding separation of physical and logical information flows, see *NIST SP 800-53* (2013), p. F–18.

exception[78] may not be possible for some kinds of neuroprosthetic devices. For example, in the case of sensory neuroprosthetics receiving sensory stimuli from the environment, it may not be feasible or theoretically possible to apply filters or tests at the boundary between the external environment and the device to determine with any accuracy what the ultimate effect of the sense data may be on the psychological health and security of a device's host and thus to allow only certain information to be transmitted inward for further processing and utilization by the device and host-device system.

9. Design of devices as thin nodes

It may be difficult to implement many kinds of neuroprosthetic devices as thin nodes,[79] given the diverse range of complex tasks that such devices must perform; the multiple forms of communication and interaction that they may need to carry out with biological systems, other implanted devices, and external support systems; the high standards set for their functionality; and the fact that such devices may need to be engineered with a wide range of surplus capacities that may or may not ever be used, due to the difficulty of modifying devices to increase their capacities after their implantation in a human host. On the other hand, some kinds of passive neuroprostheses[80] may function as thin nodes if they are designed to be directly controlled by the biological processes of their human host and do not need to possess sophisticated mechanisms for the storage of digital data, wireless communication, or other functionality commonly found in mobile devices.

10. Planning of distributed processing and data storage

Some kinds of neuroprosthetic devices (such as those employing a physical neural network with holonomic or holographic storage models) may inherently utilize distributed processing and storage.[81]

11. Restricting the use of live data during system development

The use of live data during the development and testing of information systems is generally discouraged, as storing information within systems whose security functionality is not yet assured and utilizing the information in a way unprotected by an organization's existing InfoSec mechanisms and procedures creates a significant risk.[82] However, with some kinds of neuro-

---

[78] *NIST SP 800-53* (2013), p. F–189.

[79] *NIST SP 800-53* (2013), p. F–202.

[80] See Chapter One of this text for a discussion of passive neuroprosthetic devices.

[81] Regarding distributed processing and data storage, see *NIST SP 800-53* (2013), p. F–209.

[82] *NIST SP 800-53* (2013), p. F–176.

prosthetic devices it may be impossible to avoid the use of live data even during the initial development and testing phases – for example, in cases in which a neuroprosthetic device is not fully assembled in an external facility and then implanted whole into the body of a human host but is instead assembled (or, if it utilizes biological components, even 'grown'[83]) within the body of its human host, piece by piece – perhaps through the use of nano-robots[84] or other technologies. In such cases, the development process for each particular neuroprosthetic device is unique and depends on (and is guided by) the immediate feedback provided by live data generated by the cognitive or biological processes of the device's human host.

B. Memory-related controls

1. Memory protection

Traditionally, memory protection involves hardware- or software-enforced practices such as ensuring that adversaries are not able to execute code in non-executable areas of memory.[85] In the case of advanced neuroprosthetic devices, it is not only the executable memory of a device's electronic components that must be protected but also the sensory, short-term, and long-term memory of the device's human host and any memory systems that may be created by the device and host acting jointly within the host-device system.[86] For example, cyberattacks that are able to manipulate sensory memory could potentially cause the host to perform (or not perform) physical actions in a particular manner desired by an adversary, by distorting the host's understanding of his or her environment, bodily position, or other phenomena; manipulated or fabricated information contained within sensory memory would then compromise the host's short- and long-term memory after being transmitted to those systems. Directly manipulating a host's long-term memory could also cause the host to execute or not execute actions as desired

---

[83] For the possibility of neuroprosthetic devices involving biological components, see Merkel et al. (2007). For a hybrid biological-electronic interface device (or 'cultured probe') that includes a network of cultured neurons on a planar substrate, see Rutten et al., "Neural Networks on Chemically Patterned Electrode Arrays: Towards a Cultured Probe" (2007). Hybrid biological-electronic interface devices are also discussed in Stieglitz, "Restoration of Neurological Functions by Neuroprosthetic Technologies: Future Prospects and Trends towards Micro-, Nano-, and Biohybrid Systems" (2007).

[84] See Pearce, "The Biointelligence Explosion" (2012).

[85] *NIST SP 800-53* (2013), p. F–233.

[86] For experimental research with mice that suggests the possibility of eventually developing human mnemoprostheses, see Han et al. (2009) and Ramirez et al., "Creating a False Memory in the Hippocampus" (2013). For the possibility that an adversary might use a compromised neuroprosthetic device in order to alter, disrupt, or manipulate the memories of its host, see Denning et al., "Neurosecurity: Security and Privacy for Neural Devices" (2009).

by an adversary – for example, pressing a button that the host's long-term memory tells the host will have one effect, when in fact pressing the button will have a completely different effect, and the host's memory of the button's significance has been altered.

Note that there are systems and processes found (or whose existence is hypothesized) within the human mind that play roles analogous to those of the executable memory found in a traditional computer and which relate to human memory but may also involve other kinds of processes – for example, the visuospatial sketchpad described in the Working Memory model[87] or the spotlighted 'theater of consciousness' described in the Global Workspace Theory.[88]

2. Design of protections for information at rest

The phrase 'information at rest' is generally used to describe information during those times when it is physically embodied in a particular way that is seen as relatively stable – namely, it describes "the state of information when it is located on storage devices as specific components of information systems."[89] In reality, even information that is stored on physical storage devices of the most reliable and secure form imaginable is never truly 'at rest,' as the physical substrates within which information is stored (such as the ferromagnetic layer of a hard disk drive's platter) are continuously being impacted at the subatomic level by phenomena such as cosmic rays and probabilistic quantum effects, even if these phenomena rarely have impacts that are directly visible at the macroscopic level. In well-designed systems, this process of ongoing change at the subatomic level in the structure and composition of the substrates typically does not modify the contents of the information as it is accessed and interpreted by human beings; nonetheless, it has the potential to do so. The possibility that even 'information at rest' could be modified or destroyed through the impact of 'soft errors' caused by cosmic rays, other electromagnetic radiation, or random quantum effects generally increases as units of data (such as bits) are stored in smaller physical structures, such as those of a single electron.[90]

---

[87] See, e.g., Baddeley, "The episodic buffer: a new component of working memory?" (2000).

[88] See, e.g., Baars, *In the Theater of Consciousness* (1997).

[89] *NIST SP 800-53* (2013), p. F–203.

[90] For a discussion of various kinds of soft errors and approaches for preventing them or limiting their impact, see Borkar, "Designing reliable systems from unreliable components: the challenges of transistor variability and degradation" (2005); Wilkinson & Hareland, "A cautionary tale of soft errors induced by SRAM packaging materials" (2005); Srinivasan, "Modeling the cosmic-ray-induced soft-error rate in integrated circuits: an overview" (1996); and KleinOsowski et al., "Circuit design and modeling for soft errors" (2008).

For some kinds of neuroprosthetic devices that utilize biological material for storing information, new complications are added to this picture: 'information at rest' that is stored within the patterns of activity of living cells (or within DNA[91]) may be modified or destroyed over time due to the birth, growth, mutation, or death of cells or the alteration of DNA due to radiation, chemical agents, biological agents and vectors, or other factors.

C. Cryptographic protections

1. Design of cryptographic protections and keys

When attempting to secure certain kinds of neuroprosthetic devices, it may be possible (or even necessary) to develop entirely new kinds of encryption which, for example, use the unique memories or other contents of the cognitive processes of a human mind as cryptographic keys.[92]

2. Planning of cryptographic key management

The need to maintain possession and confidentiality of and access to cryptographic keys[93] that are necessary for the effective functioning of a neuroprosthetic device becomes even more critical if failure or unauthorized use of the device has the potential to cause physical or psychological harm to the device's user or others.[94] The **escrowing** of encryption keys may be a necessary practice but also one that must be carried our carefully – especially if a neuroprosthetic device contains components dependent on the encryption key which, due to their implantation in the host's body, cannot easily be updated, otherwise modified, or replaced if the key should be lost or disclosed to unauthorized parties.

---

[91] For a discussion of the possibilities of using DNA as a mechanism for the storage of data, see Church et al., "Next-generation digital information storage in DNA" (2012).

[92] For such possibilities, see Thorpe et al., "Pass-thoughts: authenticating with our minds" (2005); Mizraji et al., "Dynamic Searching in the Brain" (2009), where the term 'password' is used in a more metaphorical sense than the typical meaning in information security, although the dynamic memory searching mechanisms described there could potentially also serve as the basis for an authentication system; and Gladden, "Cryptocurrency with a Conscience: Using Artificial Intelligence to Develop Money that Advances Human Ethical Values" (2015). Regarding cryptographic protections, see NIST SP 800-53 (2013), p. F–196.

[93] NIST SP 800-53 (2013), p. F–195.

[94] See Chapter Three of this text for proposed approaches to storing the cryptographic key for an implanted neuroprosthetic device on the host's body in the form of an external token, bracelet, tattoo, or other item, in order to provide device access to medical personnel in the case of a medical emergency affecting the device's host.

3. Full-device encryption

Although desirable from an InfoSec perspective, full-device encryption and container-based encryption[95] may not be possible for the contents of some kinds of neuroprosthetic devices, such as those storing information in a physical neural network.

4. Planning encryption of outgoing device transmissions

Encrypting outgoing transmissions[96] may be impossible, for example, in the case of neuroprosthetic devices that transmit information in the form of electrochemical signals that must be interpretable by natural biological neurons within the body of a device's host; in such cases, a device may be required for functional and operational reasons to transmit information in a form that can be received and processed by the biological and psychological systems of the device's host, regardless of whether that form is naturally secure. At the same time, some devices (e.g., mnemoprostheses that are fully integrated into a natural holographic storage system of the human brain) that store and transmit information in a form that can only be processed and interpreted by the mind of the human host in whom the devices are implanted may enjoy a natural (if unconventional) form of encryption.

D. Device power and shutoff mechanisms

1. Design of device power supply and cabling

Providing an adequate and reliable power supply[97] that is protected against intentional or unintentional damage or destruction is a major challenge for the designers and operators of advanced neuroprostheses. Some devices may be able to draw on natural power sources that are present in (or can be provided through) the natural biological systems of their human host. Such 'energy harvesting' systems for implantable devices already exist. Some gather energy from sources such as body heat or the kinetic energy resulting from movement of their host's body and can often produce more than 10 milliwatts of power.[98] Other systems utilize implantable enzyme-based biofuel cells that are able to generate power from substances such as glucose and oxygen found in the host's body.[99] There are significant practical constraints on the amount of power that can be obtained from such sources.

---

[95] *NIST SP 800-53* (2013), p. F–31.

[96] *NIST SP 800-53* (2013), p. F–193.

[97] *NIST SP 800-53* (2013), p. F–133.

[98] See Mitcheson, "Energy harvesting for human wearable and implantable bio-sensors" (2010).

[99] See Zebda et al., "Single glucose biofuel cells implanted in rats power electronic devices" (2013), and MacVitte et al., "From 'cyborg' lobsters to a pacemaker powered by implantable biofuel cells"

Other neuroprostheses may be passive devices that rely on the activity of a host's natural neurons or other biological structures or systems to control and manipulate a device and which thus do not need their own power source. Some kinds of nanorobotic swarms may be able to draw power from chemicals found (naturally or through artificial addition) within the bloodstream of their human host. Other devices may be able to receive electricity provided wirelessly, such as through radio frequency induction.[100] Other devices may require periodic recharging through connection of a physical power cable to an external power port, permanent connection of such a cable, or the periodic replacement of a battery by means of some cover or port that is accessible either on the surface of a host's body or through an invasive surgical procedure.

2. Design of emergency shutoff mechanisms for devices and systems

For many kinds of general-purpose computers used within organizations, the recommended best practice is for a computer to include an emergency shutoff switch that can be easily accessed and used by authorized personnel, should the need arise – but which cannot be accessed or used by unauthorized parties.[101] In the case of advanced neuroprosthetic devices, a number of factors will influence whether a particular device should include a physical emergency shutoff switch and, if such a switch does exist, who will have access and authorization to use it. In some cases, the presence and use of a physical emergency shutoff switch that can shut off power to a neuroprosthetic device could cause permanent physical or psychological harm to the device's host or to others; in other situations, the presence and use of such a shutoff switch may be needed precisely in order to prevent such harm. In some cases, it is essential that the host of a neuroprosthetic device have access to such a shutoff switch (because he or she will be best positioned to know when it should be used and to physically activate it), while – insofar as possible – other persons in the host's vicinity should be prevented from knowing about the shutoff switch's existence or being able to access and use it. In other cases, the kind of emergency situations that would require immediate use of the shutoff switch would also render the device's human host physically or psychologically incapable of utilizing the switch; in these cases, the shutoff switch should be physically accessible to bystanders and other persons, and it may even be desirable to install a light or audible alarm or other system to

---

(2013).

[100] See Borton et al., "Implantable Wireless Cortical Recording Device for Primates" (2009).

[101] Regarding emergency shutoff methods, see *NIST SP 800-53* (2013), p. F-133.

catch the attention of emergency medical personnel or other bystanders if the device detects a situation that calls for the use of the shutoff switch.[102]

E. Program execution protections

1. Design of protected environments for code execution

Software whose source code is unavailable to an organization or which is suspected of containing malicious code is often installed and executed by an organization only within protected and physically or virtually isolated machines running with minimal privileges.[103] For some kinds of neuroprosthetic devices whose functionality and behavioral characteristics are inherently highly influenced by or dependent on the biological structures or processes of their human host, it may not be possible to construct protected environments that fully replicate the functioning of such devices while remaining physically or virtually segregated from an actual human host. In such cases, software may need to be run within its live production environment, if it is to be run at all.

On the other hand, advanced neuroprosthetic devices may also create entirely new possibilities for constructing protected and physically or virtually isolated environments in which potentially malicious code can be run, insofar as they may allow InfoSec personnel possessing sensorimotor neuroprostheses to create and interact with information systems in a virtual environment that is separated from physical organizational systems.

2. Use of a non-modifiable operating system and applications

Designing an implantable neuroprosthetic device in such a way it loads and runs its operating system and applications from a storage medium that is permanently embedded within the device and which is hardware-enforced as read-only may be desirable,[104] insofar as it helps ensures that the device's operating system and environment will not be illicitly altered or compromised by an adversary's modification of the stored programs. However, the implications of such a practice must be carefully weighed. For example, it may sometimes occur that the operating system or applications contained on a device's read-only storage medium may need to be updated or upgraded in order to address vulnerabilities in the implanted versions of the programs that have become known. It may be difficult to implement such updates in

---

[102] See Chapter Three of this text for a discussion of emergency access to implanted neuroprosthetic devices and, in particular, the possible use of subcutaneous buttons.

[103] *NIST SP 800-53* (2013), p. F–227.

[104] See *NIST SP 800-53* (2013), pp. F–207–08.

situations in which it is not legally, ethically, or practically simple for a device's operator to remove or physically access the device in order to replace or alter the storage medium after the device's implantation.

3. Planning the role of platform-independent applications

Typically, platforms are understood as "combinations of hardware and software used to run software applications. Platforms include: (i) operating systems; (ii) the underlying computer architectures, or (iii) both."[105] The concept of a 'platform' may take on new meanings in the context of implantable neuroprosthetic devices. In some cases, the relevant 'platform' may comprise an implantable mobile computer that possesses a conventional architecture and runs a common operating system like Windows, Android, or Linux. In other cases, the 'platform' may consist of an electronic device in the form of a physical neural network comprising millions or billions of artificial neurons that are not capable of running an operating system or executing 'programs' as traditionally understood but which may nonetheless be taught to perform certain complex patterns of behavior. In other cases, the platform may include a passive device composed of biological material or electronic components that are directly guided and controlled by the activity of the cognitive and biological processes of a device's human host; in this situation, the neuroprosthetic device provides the hardware but the platform's software is found in the body or mind of its human host. This highlights the possibility that in some cases, it may not be possible to identify or understand the 'platform' created by a neuroprosthetic device simply by referring to the synthetic device itself; the platform may be constituted by or found within the larger host-device system as a whole.

It is often beneficial to utilize applications that can run on multiple platforms, insofar as this enhances application portability and the possibility of running key applications on alternate platforms, in the case of some emergency that renders their primarily platforms compromised or unavailable.[106] However, in the case of some kinds of advanced neuroprostheses, it may not only be true that applications designed for one *type* of neuroprosthetic device will be unable to run on other types of neuroprostheses, but even that applications developed for use on one neuroprosthesis implanted within a particular human being may be unable to run on other devices of the *same type* that are implanted in other human beings. Some neuroprosthetic devices may potentially store application information in biological material that incorporates the DNA of a device's human host and cannot be utilized with other

---

[105] *NIST SP 800-53* (2013), p. F–203.
[106] *NIST SP 800-53* (2013), p. F–203.

human hosts; other applications may be customized to interface with the unique physical structure or cognitive processes found in the natural biological neural network of a particular human host and thus will not function if run on another person's device.[107]

4. Protecting boot firmware

Some kinds of passive neuroprosthetic devices that are directly controlled by the 'operating system' provided by the biological structures and processes of their host's brain – as well as devices that include a physical neural network and whose functionality grows organically over time through learning and training – may not possess boot firmware as it is traditionally understood.[108]

5. Protections against the introduction or manipulation of binary or machine-executable code

For some kinds of neuroprosthetic devices (e.g., those utilizing a physical neural network of biomimetic synthetic neurons), certain types of biochemical or electrochemical stimuli allowed to reach a device's synthetic neurons could constitute a form of 'machine-executable code,' if the stimuli cause the neurons or their connected systems to respond by executing particular behaviors.[109]

6. Procedures for authentication of remote commands

It is especially important for a neuroprosthetic device to properly authenticate remote commands[110] in cases in which the device receives instructions from external medical control or support systems that can affect or determine the device's impact on critical health functions of its human host.

7. Controls on the execution of mobile code

By virtue of their highly customized design and structure, many neuroprosthetic devices may be incapable of using common mobile code technologies (such as JavaScript or Flash animations).[111] Nevertheless, it is important that the designers of neuroprostheses and developers of their operating sys-

---

[107] See the device ontologies in Chapters One and Two of Gladden (2017) for possible ways in which a neuroprosthetic device may be customized for the unique biological structures and processes – potentially as reflected in the unique psychological characteristics or knowledge – of a particular human host.

[108] See Chapter One of this text for a discussion of passive neuroprostheses of this sort. Regarding boot firmware, see *NIST SP 800-53* (2013), pp. F–226-27.

[109] Regarding binary and machine-executable code, see *NIST SP 800-53* (2013), p. F–227.

[110] *NIST SP 800-53* (2013), p. F–219.

[111] *NIST SP 800-53* (2013), p. F–198.

tems implement adequate controls to account for the possibility that operators or hosts may attempt to install mobile code on the devices or that, for
example, websites visited using a web browser or other software on a neuroprosthetic device might attempt to download and execute such code on the
device.

8. Process isolation

The use of traditional practices such as hardware separation and thread
isolation[112] may not be possible in the case of neuroprosthetic devices which,
for example, utilize a physical neural network for storing and processing data.

F. Input controls

1. Input validation procedures

Information input validation is used to protect systems from being compromised through attacks that target the structured messages that are frequently used by an information system for communications between its different components or subsystems and which may include a combination of
control information, metadata, and raw, unstructured contents.[113] If information input is not properly validated through adequate prescreening and
filtering of raw input from external systems or agents, it is possible that an
adversary could supply carefully designed raw input to one component of a
system that would then include the raw input in a structured message sent to
a different component that might erroneously interpret the raw input as
though it were control information or metadata. With some kinds of sensory
or cognitive neuroprostheses, for example, there may exist a theoretical possibility that simply by presenting certain carefully crafted forms and patterns
of environmental stimuli in such a way that they can be absorbed as raw input
by a host's sensory organs (e.g., perhaps by generating a particular series of
tones that can be detected by a host's natural ears or auditory neuroprostheses, displaying certain text or symbols on a monitor viewed by the host's natural or artificial eyes, writing particular sequences of code as graffiti on the
side of a building that the host will see, or uttering a particular string of words
to the host in conversation), such raw input will be passed along to other
components within the host's neuroprosthetic device or host-device system
through a structured communication in such a way that the raw input would
be interpreted as metadata or control information that will be executed or

---

[112] *NIST SP 800-53* (2013), pp. F–210–11.
[113] *NIST SP 800-53* (2013), p. F–229.

otherwise utilized by the neuroprosthesis to generate some action or behavior desired by an adversary.[114]

In addition to the purely technical kind of information input validation needed to prevent such occurrences, the host or operator of a neuroprosthetic device may also potentially use such mechanisms for the prescreening and filtering of raw input (e.g., stimuli from the external environment detectable by sensory organs) to screen out particular kinds of content which the host or operator might find objectionable or undesirable on other grounds – whether for legal, ethical, cultural, or aesthetic reasons or because the blocked or limited types of content have a negative operational impact on the functionality of the device or other biological or synthetic systems or processes within its host.

2. Controls on embedded data types

In a similar fashion, controls may need to be implemented to ensure, for example, that sense data being received from the external environment by an artificial sensory organ does not contain embedded patterns of data that would be detected and interpreted by the device as (potentially malicious) executable code.[115]

3. Formulation of security policy filters

Security policy filters[116] may be implemented in order to filter, for example, the kinds of auditory sense data that are permanently recorded by an artificial ear not because the device itself is technologically incapable of recording certain kinds of information but because it should not be permanently recorded due to information security considerations.

---

[114] Hansen and Hansen discuss the hypothetical case of a poorly designed prosthetic eye whose internal computer can be disabled if the eye is presented with a particular pattern of flashing lights; see Hansen & Hansen, "A Taxonomy of Vulnerabilities in Implantable Medical Devices" (2010). Although that example is of a different sort than the hypothetical cases just presented here – insofar as the case presented by Hansen and Hansen might conceivably involve a purely physical flaw or other vulnerability in the prosthetic eye that does not involve raw data being interpreted as structured data or metadata – it reflects the same basic notion that the functioning of a neuroprosthesis could be disrupted or manipulated by providing the device with certain kinds of raw data.

[115] Regarding controls on embedded data types more generally, see *NIST SP 800-53* (2013), p. F–15.

[116] *NIST SP 800-53* (2013), p. F–16.

G. Design of a logical access control architecture

1. Planning of mandatory access controls

In some cases it may be inappropriate and potentially unethical and illegal to implement mandatory (non-discretionary) access controls[117] which, for example, prevent a device's human host from granting others access to information stored in or generated with the aid of the device. For example, imagine a neuroprosthesis that enhances its human host's powers of imagination;[118] if the end-user license agreement acknowledges that the host is the sole owner of all intellectual property (such as thoughts and ideas) that are generated with the aid of the device, the device should arguably not include controls that attempt to place mandatory limits on the user's ability to share that property with others and which block the user from utilizing his or her discretion in extending access rights to others.

2. Designing for least privilege

'Least privilege'[119] may have a unique meaning in the case of some advanced neuroprostheses whose human hosts are legally and ethically expected to possess full privileges for all aspects of a device's operation and who may determine – not during the development stage of the device but only after its implementation – how to assign privileges to other parties, subject to regular unilateral modification according to the host's wishes.

3. Isolation of access- and flow- control functions

Security functions that can (and ideally should) be segregated from the access- and flow-control enforcement functions built into a device include "auditing, intrusion detection, and anti-virus functions."[120] In the case of some neuroprostheses (such as those utilizing a physical neural network), it may not be possible to isolate such functions if they are both stored and executed holographically by components that execute many of a device's functions.

---

[117] *NIST SP 800-53* (2013), p. F–11.

[118] For discussion of such possibilities, see Cosgrove, "Session 6: Neuroscience, brain, and behavior V: Deep brain stimulation" (2004); Gasson, "Human ICT Implants: From Restorative Application to Human Enhancement" (2012); and Gladden, "Neural Implants as Gateways to Digital-Physical Ecosystems and Posthuman Socioeconomic Interaction" (2016).

[119] *NIST SP 800-53* (2013), p. F–179.

[120] *NIST SP 800-53* (2013), p. F–185.

H. Design of authentication mechanisms

1. Determination of actions permitted without identification or authentication

Some advanced neuroprosthetic systems (e.g., those based on a physical platform utilizing nanorobots or synthetic neurons) may not be capable of carrying out user identification or authentication; in these cases, the devices may permit and perform all possible actions without identification or authentication.[121]

2. Restriction of unencrypted embedded static authenticators

In the case of neuroprostheses that store information in the form of a physical neural network, it may not be possible to force (or even enable) the system to store its information in a form that utilizes traditional encryption methods.[122]

3. Planning of device attestation

Device attestation performs "the identification and authentication of a device based on its configuration and known operating state."[123] Some neuroprosthetic devices – such as those comprising physical neural networks[124] or biological components – may not possess stable, clearly definable configurations or operating states that can be used as the basis for device attestation. However, it may be possible to perform attestation on the basis of a cryptographic hash[125] that is stored within the device or its components, even if that information is not directly utilized by the device itself in performing its normal functions.

4. Management of user identifiers

For neuroprosthetic devices that automatically run once activated, without requiring a system logon – or which simply verify that they possess an active physical and biological interface with a human host, without determining who that host is – a device may not utilize any user or administrator accounts and thus there would not be unique account identifiers.[126] In other

---

[121] Regarding the InfoSec implications of actions permitted without identification or authentication, see *NIST SP 800-53* (2013), p. F–24-25.

[122] Regarding the encryption of embedded static authenticators, see *NIST SP 800-53* (2013), pp. F–97-98.

[123] *NIST SP 800-53* (2013), p. F–94.

[124] See the device ontology in Chapter One of Gladden (2017) for a discussion of such devices.

[125] *NIST SP 800-53* (2013), p. F–94.

[126] See Chapter Three of this text and its discussion of biometrics for the possibility that a neuroprosthetic device might detect whether it is situated within a living human being. Regarding identifier management, see *NIST SP 800-53* (2013), p. F–94.

cases, the identifier for a device's human host or operator may not be an account name or string of text as commonly used but may potentially be an image, sound, electromechanical stimulus, or other kind of information that in different types of systems may generally be used for purposes of authentication rather than identification.

5. Planning authenticator management for multiple user accounts

Some implantable neuroprosthetic devices may not possess multiple 'accounts' that allow a device's operator or host to log into the system; the device may simply begin running once it is supplied with power and activated. In effect, such a device has a single account with an automatic logon. Other devices may have specialized accounts for a device's operator(s) and potentially its human host.

6. Planning of identification and authentication methods for organizational users

For neuroprosthetic devices that are acquired and operated by individual human hosts as consumer electronics devices, the robust systems for identifying and authenticating users[127] that are utilized within large institutions with dedicated IT and InfoSec personnel may not be available. On the other hand, for some kinds of neuroprostheses, a device may only physically be capable of interacting with the single human being in whose body the device is initially installed – thereby eliminating both the need and ability to create multiple user accounts or distinguish between organizational and non-organizational users.

7. Planning acceptance of third-party credentials

Allowing the use of third-party credentials to authenticate non-organizational users[128] of a neuroprosthetic device may be one approach to addressing the fact that, for example, the human host of a neuroprosthesis might experience a medical emergency when he or she is in a public place or otherwise unable to rely on specialized medical support services provided by his or her employer or healthcare provider. In such a circumstance, emergency medical responders who have no previous association with a device's host or operator may need to acquire immediate full access to the device and its functionality and an ability to override existing settings and control its operation in order to provide medical treatment and avoid harm to the host or others. It may be possible for local, national, or international governmental agencies, licensing and certification bodies, or associations of licensed medical personnel or other first responders to serve as third parties issuing credentials to individual personnel which the designers, manufacturers, and operators of advanced

---

[127] *NIST SP 800-53* (2013), p. F-91.
[128] *NIST SP 800-53* (2013), p. F-100.

neuroprostheses will allow their devices to accept as authenticators – either universally, or perhaps only when a neuroprosthetic device detects that its user has entered a particular biological state or is experiencing a particular medical condition.[129]

8. Architecture for adaptive identification and authentication

Some neuroprosthetic devices may utilize adaptive identification and authentication.[130] For example, the host or operator of a motor neuroprosthesis may be able to operate an artificial limb within certain nominal physical parameters without requiring special identification, but attempting to instruct the limb to operate in a way that would create a danger of significant damage to the device or its host may trigger a request from the system for additional authentication information before the instruction is executed. Similarly, for reasons of physical or psychological safety, an artificial eye or ear might possess built-in artificial constraints in the kind or quantity of incoming information that will be allowed to reach the conscious awareness of its human host; disabling such filters that limit the brightness of visual data or loudness of auditory data might be possible only after successfully submitting additional authentication information.[131]

9. Design of single sign-on capacities

If a single human host possesses multiple implanted neuroprostheses, it may be desirable for a single system (e.g., one that has direct access to the user's cognitive activity and which can be controlled by his or her thoughts) to serve as the user's interface with the collection of devices; logging on to that single gateway device would simultaneously give the user access to the other implanted systems.[132]

10. Designing password-based authentication

For some kinds of neuroprosthetic devices that interface directly with the conscious mental processes of their human host, a host's authenticator could potentially be a particular thought or memory (or the context surrounding

---

[129] For a discussion of certificate schemes, see, Chapter Three of this text and, e.g., Cho & Lee (2012), and Freudenthal et al., "Practical techniques for limiting disclosure of RF-equipped medical devices" (2007). Regarding the ability of IMDs to detect a medical emergency that is being experienced by a device's human host, see Denning et al., "Patients, pacemakers, and implantable defibrillators: Human values and security for wireless implantable medical devices" (2010), pp. 921-22.

[130] See *NIST SP 800-53* (2013), p. F–102.

[131] For a discussion of psychological, social, and cultural factors that might cause the host of an implanted device to intentionally ignore, disable, or otherwise subvert a device's security features and mechanisms – even to the extent of causing self-harm – see Denning et al. (2010).

[132] Regarding single sign-on approaches, see *NIST SP 800-53* (2013), p. F–92.

that memory) rather than a password as understood in the traditional sense of a discrete string of characters.[133] An internal thought used as a password may take on a slightly different form each time it is expressed by its user, thus it may need to be authenticated using some statistical means (perhaps employing a neural network) rather than determining whether it precisely matches some discrete piece of information used as a reference.

11. Design of authentication methods based on hardware tokens

In the case of implantable neuroprosthetic devices, it may be possible to utilize a hardware token-based authenticator that is implanted elsewhere in the body of a neuroprosthetic device's human host.[134] While the ongoing physical proximity of the hardware token to the neuroprosthetic device does not in itself guarantee that the device is still implanted within its human host, the fact that a hardware token is no longer in physical proximity to its associated neuroprosthetic device *could* indicate either that the neuroprosthetic device has been removed from its host (and should thus automatically deactivate itself and potentially wipe stored information) or that the security of the portion of the host's body in which the token was stored has been compromised (which, in some circumstances, may also be a condition that should trigger automatic deactivation of the neuroprosthesis and the wiping of information stored within it). It is also possible that implanted neuroprosthetic devices themselves could be used as authenticators to grant their host access to other (external) information systems.

12. Biometric authentication

Traditional biometric authentication methods do not require an exact match between the biometric data presented by an individual who wishes to access a system and the stored biometric data used as an authenticator; a number of both false positives and false negatives are to be expected.[135] Because of their unique (and potentially long-term or even permanent) interface with the biological structures and processes of their human host, neuro-

---

[133] Elements that could be employed in such an approach are discussed, e.g., in Thorpe et al. (2005); Mizraji et al. (2009); and Gladden, "Cryptocurrency with a Conscience" (2015). Regarding password-based authentication more generally, see *NIST SP 800-53* (2013), p. F-96-97.

[134] Regarding hardware-based authentication, see *NIST SP 800-53* (2013), p. F-98. For the use of RFID implants as authenticators, see Rotter et al., "Potential Application Areas for RFID Implants" (2012). See Chapter Three of this text for a discussion of the advantages and disadvantages of using external hardware tokens to allow medical personnel emergency access to an IMD.

[135] *NIST SP 800-53* (2013), p. F-98.

prosthetic devices may be able to utilize newly developed biometric technologies and methods that are not possible for other kinds of information systems.[136]

13. Planning of authentication feedback to users

Insofar as the process of identification and authentication might take place entirely within the cognitive processes of a neuroprosthetic device's human host, it may be possible for the device to provide full authentication feedback to the device's host (e.g., displaying the actual characters of a password that is being mentally 'typed' by the device's host, without replacing the characters with asterisks to obscure their value), without the worry that the feedback may be observed or intercepted by unauthorized parties using methods such as shoulder surfing.[137]

I. Design of session controls

1. Planning of session authenticity controls

Information systems utilize controls that protect the authenticity of sessions in order to guard against phenomena like man-in-the-middle attacks and session hijacking.[138] Some kinds of neuroprosthetic devices (such as those that possess physical neural networks and interact through ongoing synaptic communication with a human host who is also a device's operator) may not utilize sessions or other commonly employed control practices such as user accounts or authentication.

2. Restrictions on concurrent sessions

For some kinds of neuroprostheses (e.g., those that include a physical neural network that interacts directly with the memory mechanisms of their host's brain), the number of concurrent sessions[139] may be limited for technological reasons to a single session – namely, that associated with the device's human host.

3. Implementation of session lockout in response to inactivity

Automatically terminating a session after a predetermined period of inactivity[140] may be hazardous with neuroprosthetic devices whose operator and

---

[136] See Chapter Three of this text for a more in-depth investigation of unique possibilities for the use of biometrics with neuroprosthetic devices.

[137] See *NIST SP 800-53* (2013), p. F–99.

[138] *NIST SP 800-53* (2013), p. F–201.

[139] *NIST SP 800-53* (2013), p. F–23.

[140] *NIST SP 800-53* (2013), p. F–23.

host expect or require that a device always be ready to provide access and service, without the delay that would be required for reauthentication.[141]

4. Design of automatic session termination procedures

Session termination[142] may be impossible to implement for some kinds of neuroprosthetic devices. A device consisting of synthetic neurons that are fully integrated into the natural biological neural network of their host's brain may in effect run a single 'session' that will last throughout the host's remaining lifetime.

J. Wireless and remote-access protections

1. Preventing information leakage resulting from stray electromagnetic emissions

A neuroprosthetic device should be protected against "the intentional or unintentional release of information to an untrusted environment from electromagnetic signals emanations."[143] This may be especially difficult when multiple devices implanted within a single host form a body area network (BAN) or body sensor network (BSN) whose components communicate with one another through wireless signals; the use of components that transmit signals through bodily tissue using means that do not broadcast signals into the atmosphere may reduce that risk.

The danger of information leakage may also be relatively high in the case of a neuroprosthetic device implanted within the interior of its host's body that possesses no physical port or socket accessible on the external surface of the body and which must communicate with external diagnostic, control, or support systems utilizing wireless means.

2. Planning of remote access methods

A neuroprosthetic device implanted in a human host may need to remotely access or be accessed by systems[144] that belong, for example, to the device's manufacturer, operator, or a dedicated medical support team. Automated monitoring, encryption, and use of managed access control points may be desirable in such circumstances. Such access need not necessarily be wireless, if a neuroprosthetic device has an external port that allows for a wired connection.

---

[141] See Chapter Three of this text for a discussion of the need for 100% availability for some kinds of neuroprosthetic devices.

[142] *NIST SP 800-53* (2013), p. F–24.

[143] *NIST SP 800-53* (2013), p. F–138.

[144] See *NIST SP 800-53* (2013), p. F–28.

3. Protection against wireless jamming and electromagnetic interference

A system can potentially be protected from intentional jamming through the use of unpredictable wireless spread spectrum waveforms, while other technologies may provide protection against unintentional jamming or interference (e.g., from nearby devices using the same wireless frequencies).[145] This is especially important in the case of neuroprosthetic devices whose activity can have a critical impact on the health of their human host and whose successful functioning depends on effective wireless communication with other implanted devices or external support systems.

K. Design of backup capabilities

1. Planning of alternate data processing site(s)

The use of alternate processing sites[146] for the processing of information by a neuroprosthetically augmented information system may not be possible if the act of processing is in part performed by the neurons within the host's brain or other biological systems within the host's body or if processing can only be carried out by a device when it enjoys a direct physical interface with the host's brain or body.

2. Planning of alternate data storage site(s)

The use of an alternate storage site[147] external to the body of a device's human host for storing information generated by an implanted neuroprosthetic device may not be possible for some devices that store information in particular kinds of systems (such as a physical neural network[148]) or which lack an adequate means of transmitting the relevant quantity and type of information to external systems.

3. Design of backup communications systems

A neuroprosthetic device may or may not be capable of using general-purpose communications technologies and services as a backup system if the device's own telecommunications system (which may be proprietary or demonstrate unique specifications for its speed, capacity, and format) were to fail or be disrupted.[149]

---

[145] *NIST SP 800-53* (2013), p. F-211.

[146] *NIST SP 800-53* (2013), pp. F-83-84.

[147] *NIST SP 800-53* (2013), p. F-83.

[148] See the device ontology in Chapter One of Gladden (2017) for a discussion of the structure and mechanics of such systems that include or comprise physical artificial neural networks.

[149] Regarding contingency planning for telecommunications services, see *NIST SP 800-53* (2013), p. F-85.

4. Design of information backup methods

For some kinds of advanced neuroprostheses (such as those utilizing a complex physical neural network and holographic storage system) it may be impossible to create a coherent backup, insofar as this would require taking a 'snapshot' of the entire constantly-changing system at a single instant, but the processes available for detecting and recording the state of information within the system's components can only scan components sequentially and require a long period of time to complete a single full scan of the system.[150]

5. Planning of safe mode behavior for devices

For some kinds of neuroprosthetic devices it may be desirable to develop a safe mode[151] with a predefined and limited set of features and operations that can either be manually activated by a device's operator or human host if it becomes apparent that the host is entering (or about to enter) some situation in which unrestricted operation of the device would be hazardous to the host or others or which will be automatically activated if the device detects that certain conditions are met.[152] Note that if activating a device's safe mode will result in a loss of consciousness or in some other impairment for the device's host, then the device may also need to possess a mechanism for determining when to automatically exit safe mode and resume normal operations, insofar as the host would not be able to manually initiate such an action.

L. Component protections

1. Controls to assure component authenticity

Preventing the use of counterfeit components is especially important in the case of neuroprosthetic devices in which the discovery that counterfeit components (which may potentially be constructed from toxic materials or of otherwise substandard quality) had been used in a neuroprosthetic device

---

[150] Regarding related technologies that have been proposed by some transhumanists as a possible path toward brain emulation of 'mind uploading,' see Koene, "Embracing Competitive Balance: The Case for Substrate-Independent Minds and Whole Brain Emulation" (2012); Proudfoot, "Software Immortals: Science or Faith?" (2012); Pearce (2012); Hanson, "If uploads come first: The crack of a future dawn" (1994); and Moravec, *Mind Children: The Future of Robot and Human Intelligence* (1990). Regarding information system backups, see *NIST SP 800-53* (2013), p. F–87.

[151] *NIST SP 800-53* (2013), p. F–89.

[152] Regarding the possibility that an IMD could discern when, e.g., a medical emergency is being experienced by its human host, see Denning et al. (2010), pp. 921-22. See Chapter Three of this text for a broader discussion of failure modes for neuroprosthetic devices during emergency situations.

may necessitate complex, expensive, dangerous, and legally and ethically fraught surgery to extract a device and replace the components.[153]

2. Customized design for critical device components

For some kinds of neuroprosthetic devices, the in-house development of customized, nonstandard components (which may be less vulnerable to standard attacks that adversaries might be likely to employ[154]) may be a natural and even necessary aspect of a device's development. For example, in the case of neuroprostheses that utilize biological components that incorporate a host's DNA or whose security functionality depends on unique features of the host's mind (such as memories unique to that host[155]), each device may in effect be deeply customized and 'nonstandard.'

3. Designing approaches to device identity and traceability

For neuroprostheses that are housed permanently within the body of a human host and that cannot easily be physically inspected or extracted, the confirmation of identity and traceability of such devices and their components may need to be accomplished using technologies such as RFID tags[156] that can be checked wirelessly by a reader external to a host's body. In the case of some kinds of advanced neuroprostheses utilizing biological components, it may be possible to incorporate identifying marks, codes, supporting documentation, and other information into the genetic sequences of the biological material.[157]

4. Planning tamper-resistance mechanisms for the entire SDLC

The tamper-resistance mechanisms that are legally and ethically permissible and practically feasible during the pre-implantation production and testing of a neuroprosthetic device may be entirely different from those that are possible and desirable during the operations and maintenance or disposal phases of the device's SDLC.[158]

---

[153] For medical risks relating to surgery for the implantation of even 'simple' implants such as passive RFID devices, see Rotter et al., "Passive Human ICT Implants: Risks and Possible Solutions" (2012). Regarding component authenticity, see *NIST SP 800-53* (2013), p. F–180.

[154] See *NIST SP 800-53* (2013), p. F–181.

[155] See Chapter Three of this text for a discussion of the possibility of using a host's thoughts and memories as biometric access controls.

[156] Regarding identity and traceability, see *NIST SP 800-53* (2013), p. F–172.

[157] For such possibilities, see Church et al. (2012).

[158] Regarding the development of tamper-resistance mechanisms for multiple phases of the SDLC, see *NIST SP 800-53* (2013), p. F–180.

5. Removal of unsupported system components

Typical best practices of removing and replacing system components[159] once they are no longer supported by their designer, manufacturer, or provider may be difficult or impossible to implement in the case of devices that have been implanted in a human host and whose removal would require complex, expensive, or dangerous surgical procedures or would otherwise create a possibility of physical or psychological harm for a device's host.

M. Controls on external developers and suppliers

1. OPSEC activities targeted at device or component suppliers

An organization may employ utilize operations security practices and safeguards in relation to current and potential suppliers.[160] In that context,

> OPSEC is a process of identifying critical information and subsequently analyzing friendly actions attendant to operations and other activities to: (i) identify those actions that can be observed by potential adversaries; (ii) determine indicators that adversaries might obtain that could be interpreted or pieced together to derive critical information in sufficient time to cause harm to organizations; (iii) implement safeguards or countermeasures to eliminate or reduce to an acceptable level, exploitable vulnerabilities; and (iv) consider how aggregated information may compromise the confidentiality of users or uses of the supply chain.[161]

As employed by some organizations, OPSEC practices and tactics may sometimes lead an organization to "withhold critical mission/business information from suppliers and may include the use of intermediaries to hide the end user, or users, of information systems, system components, or information system services."[162] Before they can be implemented, careful attention must be given to the legal and ethical implications of such OPSEC practices in the case of advanced neuroprosthetic devices. For example, a supplier that is led to believe that it is producing components for use in neuroprosthetic devices to be implanted in mice for experimental research may utilize a different level of care in producing the components (and, indeed, may make a different decision about whether to enter into a contract to supply the components) than it would have done had it been aware of the fact that its components would be incorporated into neuroprostheses to be implanted in human hosts for use in their performance of critical tasks – even if the supplier's knowledge of the true circumstances of the devices' ultimate use would not in any way have affected the specifications of the components that the client

---

[159] *NIST SP 800-53* (2013), p. F–182.

[160] *NIST SP 800-53* (2013), p. F–171.

[161] *NIST SP 800-53* (2013), p. F–171.

[162] *NIST SP 800-53* (2013), p. F–171.

organization had asked the supplier to produce. Complex problems involving legal liability, moral responsibility, corporate social responsibility, and informed decision-making can arise if OPSEC activities prevent the free and robust flow of accurate information between an organization, its suppliers, and other parties involved with the development and implementation of advanced neuroprosthetic devices.

On the other hand, in situations in which the personal health information and sensitive data about the cognitive processes (including thoughts, memories, and emotions) of particular human hosts is involved, an organization may have a legal and ethical responsibility not only to conceal the detailed information that is received, stored, generated, or transmitted by the neuroprosthetic devices that it operates but even any incidental or circumstantial information that could potentially be used by suppliers to ascertain (or even guess at) the identity of a device's human host.[163] This may be especially important in cases in which a host is a significant political, business, artistic, or entertainment figure, a military or police operative, or some other individual whom unauthorized parties may have a particular interest in observing, stealing information from, blackmailing, extorting, or otherwise compromising or exploiting (i.e., through so-called 'whaling' attacks).

2. Formulation of procedures, standards, and tools for developers

With advanced neuroprosthetic devices, it is especially important that suppliers and developers follow a well-defined and thoroughly documented development process for components and services, since in the case of unforeseen operational emergencies (such as a critical negative impact on the health of a device's human host that arises unexpectedly) it may be necessary to quickly retrace and analyze steps in the development of a component or service in order to formulate a response that can prevent serious harm to the device's host or to others.[164]

3. Security testing policies for third-party developers

Some practices that an organization may typically require of third-party software providers – such as static code analysis and manual code analysis[165]

---

[163] For discussions of such issues, see Kosta & Bowman (2012); McGee (2008); Shoniregun et al., "Introduction to E-Healthcare Information Security" (2010); Hildebrandt & Anrig (2012); and Brey (2007).

[164] Regarding the creation of development processes, standards, and tools, see *NIST SP 800-53* (2013), p. F–174.

[165] *NIST SP 800-53* (2013), pp. F–166-69.

may not be relevant or possible in cases in which, for example, neuroprosthetic devices include physical neural networks that do not execute programs or code as traditionally understood.[166]

4. Supervision of developer configuration management

Effective monitoring and supervision of developers' configuration management[167] by an organization is especially important in cases where the developer of, for example, the OS or software applications installed in implanted neuroprosthetic devices maintains direct access to the software and periodically pushes out software updates, patches, or configuration changes to devices that are implanted and in use.[168]

5. Protections for component supply chains

It is possible that adversaries may choose to identify and target an organization's supply chain of components or services needed for the design, production, implementation, maintenance, or operation of organizational information systems rather than directly targeting the information systems themselves. *NIST SP 800-53* thus notes that "Supply chain risk is part of the advanced persistent threat (APT)"[169] that organizations face. An adversary could potentially execute such an attack by compromising a supplier and covertly corrupting or manipulating the supplier's production processes, so that components produced by the supplier for an organization have been produced using improper materials that will disintegrate, break, or otherwise fail (or, in the case of an advanced neuroprosthesis, potentially poison a device's host or release other biologically or psychologically active agents into the host's body) after the information system is in use, or components may have been corrupted with malware or designed with unauthorized backdoors that will allow adversaries unauthorized access to the system after it is in use.[170] In the case of some advanced neuroprosthetic devices, the number of suppliers producing certain necessary components may (at least initially) be quite small: such a phenomenon is disadvantageous, insofar as it bars an organization

---

[166] For a discussion of, e.g., neuroprosthetic devices based on physical neural networks that do not execute traditional programs, see the device ontology in Chapter One of Gladden (2017).

[167] *NIST SP 800-53* (2013), pp. F–164-66.

[168] See Chapter Five of this text for a discussion of the roles and responsibilities of OS and application developers for neuroprosthetic devices.

[169] *NIST SP 800-53* (2013), p. F–170.

[170] Regarding backdoors intentionally built into implantable medical devices to allow emergency access to medical personnel – which could potentially be exploited by sufficiently knowledgeable adversaries – see Clark & Fu, "Recent Results in Computer Security for Medical Devices" (2012); Halperin et al., "Security and privacy for implantable medical devices" (2008); and Chapter Three of this text.

from pursuing the typical approach of minimizing supply chain risk by acquiring components from multiple suppliers, although it is potentially advantageous, insofar as it allows an organization to concentrate its information security resources and efforts on securing the operations of just a single supplier or small group of suppliers.

6. Scrutiny of external information system services and providers

When engaging external information system service providers[171] in relation to advanced neuroprostheses, an organization must be careful that the external service providers do not receive access to the biological processes of devices' human hosts in a way that may be illegal or unethical; consent that has been given by the hosts for the organization to access and use information or to manipulate their internal biological or cognitive processes may or may not apply to external service providers acting on behalf of the organization. Moreover, it is not enough for an organization to satisfy itself that it does not possess conflicts of interest or other potentially harmful characteristics that could impair or call into question its ability to ensure the information security of devices' human hosts; an organization should also seek and obtain assurance that potential external service providers do not possess conflicts of interest, ulterior motives, or other traits that may give reasons for neglecting or actively compromising the information security of neuroprosthetic devices' human hosts, either as a group or in specific cases (e.g., with regard to human hosts who are significant political, military, business, or entertainment figures or otherwise likely targets of whaling attacks).

SDLC stage 3: device deployment in the host-device system and broader supersystem

The third stage in the system development life cycle includes the activities surrounding deployment of a neuroprosthetic device in its human host (with whom it forms a biocybernetic host-device system) and the surrounding organizational environment or supersystem. The development or implementation of security controls in this stage of the SDLC is typically performed by a device's operator with the active or passive participation of its human host. Such controls are considered below.

A. Environmental protections

1. Fire protection methods

Although it may not be possible to build full fire-suppression systems directly into neuroprosthetic devices, some devices that are composed of flammable materials or whose operation has the potential to generate excessive

---

[171] *NIST SP 800-53* (2013), pp. F–162-64.

heat or sparks may need to at least include built-in fire-detection systems, with a device's host or operators maintaining external fire-suppression systems that are always available for use in emergencies.[172]

2. Design of temperature and humidity controls

It may be crucial to implement systems that maintain a neuroprosthetic device within a predetermined range of temperatures and which ensure that other internal and external environmental conditions are maintained – not only to ensure that the device remains within appropriate operating parameters but also that surrounding biological tissue and processes (which may be sensitive to even minute temperature changes) are not damaged by excessive heat or other emissions from the unit.[173]

3. Planning the location of information system elements

General best practices include choosing – insofar as is feasible – to house information system components in a location that is protected against "flooding, fire, tornados, earthquakes, hurricanes, acts of terrorism, vandalism, electromagnetic pulse, electrical interference, and other forms of incoming electromagnetic radiation" and which lacks "physical entry points where unauthorized individuals, while not being granted access, might nonetheless be in close proximity to information systems and therefore increase the potential for unauthorized access to organizational communications (e.g., through the use of wireless sniffers or microphones)."[174] In the case of advanced neuroprosthetic devices, the fact that a device's human host is potentially able to take a device anywhere in the world may make it difficult or impossible to prevent the neuroprosthesis from being brought into areas exposed to such situations – a danger that is present for other types of mobile devices more generally.

4. Design of emergency power systems

For some kinds of neuroprosthetic devices, it may be possible and desirable to utilize an emergency power system[175] (such as one that requires an external power cable to be plugged into a visible external port or jack in a neuroprosthetic device) that is not practical in non-emergency situations, when a device's host expects to be able to move freely at will between different environments without the need to be plugged into a fixed power source. Small internal batteries may also be able to provide emergency power for a limited

---

[172] Regarding fire protection, see *NIST SP 800-53* (2013), p. F–135.

[173] Regarding temperature and humidity controls, see *NIST SP 800-53* (2013), p. F–135.

[174] *NIST SP 800-53* (2013), p. F–137.

[175] *NIST SP 800-53* (2013), p. F–134.

period of time, even if they would be inadequate for powering a device for sufficiently long periods during everyday non-emergency use.

B. Contingency planning

1. Development of contingency plans

The development of effective contingency plans[176] for advanced neuro-prosthetic devices is essential, insofar as a failure or disruption in service for some devices may instantaneously result in life-threatening harm for a device's human host or others. A contingency plan may contain procedures for continuing or resuming either all or some of a device's functions and preserving critical assets in the face of various disruptions – if not through the device itself then through other available systems.

2. Contingency training

Contingency training[177] may be especially important for a device's human host if the failure or disruption of the device's service will leave the host with only a limited window of time in which to carry out critical remedial actions before the service failure leaves the host incapacitated and unable to carry out such steps.

3. Testing of contingency plans

Some kinds of contingency plan testing (such as the use of walk-throughs and checklists)[178] may be easy to carry out; however, an accurate full-scale simulation of some kinds of contingencies may be difficult to perform insofar as it would require simulating certain kinds of mental phenomena or incapacities on the part of a human host and the impact that these would have on the host, and replicating such conditions cannot be accomplished without causing actual harm to the host.

C. Tracking of system component inventory

A neuroprosthetic system's operator should ideally keep an inventory of all devices in use that records each device's "manufacturer, device type, model, serial number, and physical location."[179] However, in the case of some neuroprosthetic devices (such as those grown or assembled from living biological material or comprising a swarm of myriad nanorobots) it may be dif-

---

[176] *NIST SP 800-53* (2013), p. F–78.

[177] *NIST SP 800-53* (2013), p. F–81.

[178] *NIST SP 800-53* (2013), p. F–82.

[179] *NIST SP 800-53* (2013), p. F–73.

ficult to adequately capture the nature of a particular device using such descriptors, as each device may, in effect, be wholly unique and possess a form that is constantly shifting and evolving.[180]

D. Selection of device recipients and authorization of access

In the case of neuroprostheses operated by large institutions (e.g., devices operated by a military or intelligence agency for intelligence-gathering purposes), an organization may maintain comprehensive and detailed records of which human beings are serving as the hosts for which devices; no other human beings are authorized to serve as the hosts for those devices, and the list of other individuals (e.g., organizational medical personnel) who are authorized to gain physical access to the devices is limited. On the other hand, with neuroprosthetic devices that are sold to the public through retail outlets as consumer electronics devices, there may be no reliable centralized record of which devices are implanted in which human beings and who is a device's 'authorized' operator.[181]

E. Physical hardening of the host-device system and supersystem

1. Restrictions on the use of portable media

For computers that are permanently physically located within a secured and supervised facility belonging to an organization, it may be possible for the organization to create and enforce administrative, logical, and physical controls (such as metal cages surrounding the computers[182]) that block users from utilizing ports or slots on the computers to insert portable storage media such as flash memory cards. In the case of neuroprosthetic devices implanted in human beings who are free to travel wherever they want and who may enter diverse kinds of environments and situations, it may not be possible to implement controls that will always reliably prevent portable storage media from being inserted into a neuroprosthetic device's slots and ports; if preventing the unauthorized use of portable storage media is a top priority, it may be necessary for the device's designer and manufacturer to construct the device in such a way that no such ports or connections are present and any effort to add them by an unauthorized user would disable the device.[183]

---

[180] Such a possibility would raise challenges for use of the device ontology presented in Chapter One of Gladden (2017): in that case, the ontology would become a way of specifying a device's general structural and operational parameters rather than its exact current characteristics.

[181] Regarding physical access authorizations, see *NIST SP 800-53* (2013), pp. F–127-28.

[182] *NIST SP 800-53* (2013), p. F–124.

[183] See the device ontology in Chapter One of Gladden (2017) for different aspects of a neuroprosthetic device's physical structure and accessibility, including the presence or absence of physical input and output mechanisms.

2. Controls on access to ports and I/O mechanisms

For implantable neuroprosthetic devices which, for some reason, are required to include physical connection ports (e.g., USB or specialized proprietary ports) or input/output devices (such as microSD card readers or specialized proprietary memory chip readers) that are accessible from the exterior of their host's body, it may be desirable from the perspective of information security to disable such I/O devices at all times except for occasions when they are enabled through an explicit command from the operator of a neuroprosthesis or occasions when a medical emergency experienced by a neuroprosthetic device's host causes the ports and I/O devices to be automatically enabled in order to allow for the delivery of medical treatment by emergency personnel.[184]

Note that the legal, ethical, and practical implications of such design decisions must be carefully considered: for example, if it is commonly known that a neuroprosthetic device's exterior connection ports and I/O devices will be automatically enabled in the case of particular kinds of medical emergencies experienced by the device's human host, an adversary could potentially purposefully induce a relevant kind of medical emergency for the device's host (e.g., through a physical, biological, or chemical attack or intervention) in order to gain access to fully enabled connection ports or I/O devices that the adversary can use to compromise the neuroprosthetic device.

3. Limitations on implants' wireless transmission levels for InfoSec- and safety-related reasons

Limiting the power levels of wireless transmissions[185] from a mobile device or utilizing directional antennas is a useful practice for reasons of ensuring information security; in the case of some implantable neuroprosthetic devices it may also be desirable to help ensure the long-term health and safety of a device's host and avoid undesirable interference with other implanted systems.[186]

4. Use of lockable casings for devices

Depending on the nature of a neuroprosthetic device, it may be necessary to ensure that legitimate, licensed emergency personnel have a way to unlock

---

[184] See Chapter Three of this text for a discussion of allowing special access to neuroprosthetic devices during health emergencies that affect their human host. Regarding access to ports and I/O devices, see *NIST SP 800-53* (2013), p. F-212.

[185] *NIST SP 800-53* (2013), p. F-30.

[186] See Chapter Three of this text for a discussion of the reliance on wireless communication with external systems that is characteristic of many kinds of implantable neuroprosthetic devices.

or bypass lockable casings[187] in order to physically access a device when providing emergency medical treatment to its human host.[188]

F. Logical hardening of the host-device system and supersystem

1. Verification of transmission source and destination points

Even when a neuroprosthetic device is intended solely to transmit information between different systems that are permanently embedded within the body of its human host and not to the external environment, controls may need to be implemented to ensure that the origin and destination points of such communications are indeed the intended systems.[189]

2. Prevention of electronic discovery of devices or components

Neuroprosthetic devices that consist largely or entirely of biological material may possess a natural ability to prevent their detection as information systems (or components of such systems) by sensors or other detection mechanisms that are designed to identify, locate, and analyze conventional electronic information systems.[190]

3. Restrictions on wireless transmission strength to reduce detection potential

Even if adversaries are unable to decipher the contents of messages that are being wirelessly transmitted by a particular device, simply being able to detect the existence of the device and the fact that it is transmitting signals and to potentially pinpoint its geospatial location provides an adversary with useful information. Reducing the strength of wireless transmissions from an implanted neuroprosthetic device may reduce its detectability.[191] Some kinds of implantable neuroprosthetic devices may already be restricted to utilizing low-power transmissions in order to avoid causing potential harm or disruptive side-effects for biological systems and material within their host's body (e.g., heat generated by the absorption of radio frequency radiation).[192] At the same time, transmission signal strengths must be sufficient to ensure that

---

[187] Regarding such casings, see *NIST SP 800-53* (2013), p. F–129.

[188] For a discussion of emergency access to implantable neuroprosthetic devices, see Chapter Three of this text as well as Clark & Fu (2012); Rotter & Gasson, "Implantable Medical Devices: Privacy and Security Concerns" (2012); and Halperin et al. (2008).

[189] Regarding the related concept of domain verification, see *NIST SP 800-53* (2013), p. F–17.

[190] Regarding methods to prevent the electronic discovery of devices or components, see *NIST SP 800-53* (2013), p. F–191.

[191] Regarding methods for reducing the potential detection of wireless transmissions or devices, see *NIST SP 800-53* (2013), p. F–211.

[192] See Zamanian & Hardiman, "Electromagnetic radiation and human health: A review of sources and effects" (2005).

physical or psychological harm does not result for a device's host due to a device's failure to execute successful wireless communications with other implanted devices or external medical support or control systems.

4. Signal parameter identification to detect deceptive communications

Some information systems may utilize 'radio fingerprinting techniques' to identify and track particular devices according to the signal parameters displayed by their wireless transmissions; conversely, other devices may attempt to elicit communications from or manipulate communications with a system by intentionally imitating the signal parameters of a particular device that is already trusted by the system.[193] Utilizing appropriate techniques (such as anti-fingerprinting mechanisms employing unpredictable signal parameters) to counter such possibilities is especially important in the case of neuroprosthetic devices whose only means of communication with necessary external support and control systems is through wireless transmissions.[194]

5. Protections against spam

For individuals possessing advanced neuroprostheses that edit or replace their natural sensory input to create an experience of augmented or virtual reality,[195] spam might come in the form of messages, advertisements, alerts, or any other kind of virtual audiovisual or other sensory phenomena designed to elicit some behavior from a neuroprosthetic device's host. For individuals possessing some kinds of advanced cognitive neuroprostheses, spam may potentially even take the form of memories, emotions, desires, beliefs, or other mental phenomena that are directly inserted into or created or altered within a host's cognitive processes by some external agent.[196]

Some kinds of neuroprosthetic devices may be able to utilize spam protection mechanisms that learn what is spam by directly detecting the physical or psychological reaction presented by a device's human host to incoming messages and stimuli, thereby supplementing or enhancing traditional learning mechanisms such as Bayesian filters that are often employed in spam protection systems.[197]

---

[193] *NIST SP 800-53* (2013), p. F–212.

[194] See Chapter Three of this text for a discussion of the reliance on wireless communications that is found with many kinds of implantable neuroprosthetic devices.

[195] For the possibility that a device that has been designed to receive raw data from the external environment could have that data supplemented or replaced by other data transmitted from some external information system (which could create new opportunities for the delivery of spam content), see Koops & Leenes, "Cheating with Implants: Implications of the Hidden Information Advantage of Bionic Ears and Eyes" (2012).

[196] Regarding protections against spam, see *NIST SP 800-53* (2013), p. F–228.

[197] Regarding anti-spam systems utilizing a continuous learning capability, see *NIST SP 800-53*

6. Protections against data mining

Some kinds of neuroprosthetic devices (such as those that utilize physical neural networks and holographic storage mechanisms) may inherently possess robust protections against many typical data-mining technologies or techniques.[198]

G. Device initialization and configuration controls

1. Specification of baseline configurations

It may be difficult or impossible to specify a baseline configuration[199] for some kinds of advanced neuroprostheses, such as a mnemocybernetic device consisting of a physical artificial neural network that is integrated at the synaptic level with natural neurons in the host's brain and which does not have a set of discrete settings that can be centrally updated and applied to all individual neurons throughout the system after it has been activated. It may be impossible to intentionally roll back such a device to a previous configuration (or even to identify what such a configuration might be).

2. Automatic configuration changes

Some neuroprosthetic devices may possess a configuration that is continuously altered in automatic response to stimulation and other activity by the host's biological systems with which a device is integrated, without any means for its operator to directly control the configuration changes.[200]

3. Analysis of the InfoSec impact of configuration changes

For some neuroprosthetic devices it may not be possible to fully analyze the security impact of potential configuration changes prior to actually implementing them, if a device's exact response to the changes depends on the precise action of psychological or biological processes within the device's human host that cannot be simulated in a virtual test environment.[201]

4. Dangers of design for least functionality

In the case of a neuroprosthesis that is designed primarily to provide some necessary medical service or functional enhancement to its human host rather than to secure particular information, there may be non-security reasons

---

(2013), p. F–228.

[198] Regarding data mining protections, see *NIST SP 800-53* (2013), p. F–35.

[199] *NIST SP 800-53* (2013), p. F–64.

[200] Regarding configuration change control, see *NIST SP 800-53* (2013), p. F–66.

[201] Regarding analyses of the security impact of configuration changes, see *NIST SP 800-53* (2013), p. F–68.

for the device to offer the greatest functionality possible rather than the least allowable.[202]

5. Non-privileged access for non-security-related device functions

With some kinds of neuroprosthetic devices, it may be a technological and functional necessity for a device's operator or host to possess privileged access to the system, even when it is being used to perform non-security-related functions.[203]

6. Software usage restrictions

Software restrictions are often implemented to ensure that software is used in accordance with its licensing restrictions and to ensure that software such as a peer-to-peer file-sharing program is not used "for the unauthorized distribution, display, performance, or reproduction of copyrighted work."[204] In the case of neuroprosthetically augmented information systems, operators must be careful to ensure – for legal and ethical reasons – that any software restrictions that are capable of disabling or constraining the use of software products do not do so at a time or in a manner that could cause harm to a device's human host or others.

The policing of peer-to-peer file-sharing will also be complicated by the fact that in effect, the mind of a device's human host is a 'peer-to-peer file-sharing program' that is frequently exchanging information of all kinds with other human beings. Complex legal questions may also arise surrounding what constitutes a 'display or performance' of copyrighted material: for example, in the past, using a hidden video camera to videotape a movie that was being shown in a commercial cinema and then uploading the bootleg video to a video-streaming website would have been considered an illicit use of copyrighted material, but observing the film carefully with one's natural eyes, storing that sensory record in one's natural memory systems, and later using one's voice to describe the film to one's friends would not have been considered an illicit act. Such boundaries between licit and illicit usage may become blurred, for example, if one possesses artificial eyes or a mnemoprosthetic device that allow one to record *all* of one's daily visual experience – and not simply a film in a cinema – with high resolution and perfect fidelity or if one possesses an artificial voice-box that allows one not simply to speak with

---

[202] See Chapter Three of this text for a discussion of the trade-offs that sometimes occur between increased information security and increased functionality for neuroprosthetic devices. Regarding design for least functionality, see *NIST SP 800-53* (2013), pp. F–71-73.

[203] Regarding non-privileged access for non-security-related device functions, see *NIST SP 800-53* (2013), p. F–19.

[204] *NIST SP 800-53* (2013), p. F–76.

one's 'own' natural voice but to play back any recorded sounds, including those that one may have overheard during a film screening.[205]

Complex questions that must be resolved by law, regulation, or individual licensing agreements will also arise regarding intellectual property that is created with the aid or participation of a neuroprosthetic device. If a human host utilizes a device that enhances his or her memory, imagination, or artistic, mathematical, physical, or reasoning abilities, then any literary or artistic works, performances, inventions, or discoveries developed by the host may ultimately be the property of the neuroprosthetic device's manufacturer, provider, app developer, operator, or human host, or it may be owned jointly by some combination of parties.

7. Restrictions on the installation of software by users

Blocking the installation of software on a neuroprosthetic device by its user may or may not be legally and ethically permissible, as in some situations this may be equivalent to blocking a human being from adding thoughts, memories, or other permissible content to his or her own mind.[206]

8. Restrictions on device use

An organization may wish to prohibit or restrict the use of devices[207] that possess environmental sensing or recording capabilities (such as smartphones or cameras) within particularly sensitive facilities or areas. It is one matter for an organization to deny entry to its facility to individuals possessing handheld cameras (or to require that such individuals temporarily deposit their cameras for safekeeping with the organization's personnel upon entering the facility); it is another matter to deny entry to individuals who possess certain kinds of implantable neuroprosthetic devices, such as artificial eyes that possess the same functionality as handheld cameras. In the latter case, such neuroprostheses may, from a legal and ethical perspective, be treated as implantable medical devices, and an organization's refusal of entry or service to a person possessing such a device may in some cases be considered an unlawful form of discrimination on the basis of the person's health or medical status. Even conducting the kind of searches that may be required in order to determine the presence of some kinds of implantable neuroprosthetic devices in visitors to an organizational facility (e.g., potential custom-

---

[205] For the notion that a neuroprosthetic device could be used for sensory recording or playback, see Merkel et al. (2007); Robinett, "The consequences of fully understanding the brain" (2002); and McGee (2008), p. 217.

[206] Regarding conventional controls on user-installed software, see *NIST SP 800-53* (2013), p. F-76-77.

[207] *NIST SP 800-53* (2013), p. F-213.

ers visiting a company's retail store or showroom) may be considered an impermissibly intrusive procedure that illicitly gathers information about visitors' personal medical history and status without their express consent. Efforts by an organization to proactively jam or obstruct the functioning of some kinds of neuroprosthetic devices (such as attempts by a theater owner to prevent the use of artificial eyes to record a performance) would likely encounter legal, ethical, and practical obstacles similar to those encountered by organizations that have sought to jam the functioning of smartphones on their premises.[208]

9. Restrictions on the use of sensors and access to sensor data

An organization's ability to restrict the activation and use of environmental sensors (such as cameras, microphones, accelerometers, GPS systems, temperature gauges, and other mechanisms) in devices belonging to the organization is a critical element of information security.[209]

In the case of sensory neuroprostheses such as artificial eyes, an organization must often make a device's environmental sensing capabilities available to a device's host and operator at all times – with no delays, distortions, or other failures in service – while at the same time blocking all unauthorized parties from accessing (or potentially even knowing about the existence of) the device's sensor capacities. Adversaries who gain unauthorized access to a neuroprosthetic device's sensor capabilities could potentially use that ability to conduct covert and illicit surveillance on the device's human host, the organization by whom the host is employed, other organizations or individuals with whom the host is associated, or even organizations or individuals who have no direct connection with the host but whom the host happens to be passing by at the moment. In the case of cognitive neuroprostheses, a device itself may not possess direct access to raw sense data from the environment, but it may be able to indirectly access such data through its host's memory or other cognitive processes. Motor neuroprostheses that are used, for example, to control the movement of an artificial limb may contain accelerometers or other sensors that are intended to gather data about the position and activity of the limb but which can be utilized by adversaries to gather information about the broader physical environment, instead.

---

[208] Regarding the technological, legal, and ethical aspects of using jamming devices to block cell phone signals in places such as movie theaters and schools, see Koebler, "FCC Cracks Down on Cell Phone 'Jammers': The FCC says illegal devices that block cell phone signals could pose security risk" (2012), and Overman, "Jamming Employee Phones Illegal" (2014).

[209] *NIST SP 800-53* (2013), p. F–213.

10. Restrictions on device use outside of organizational contexts

Although it may be legally and ethically permissible and practically feasible for an organization to restrict the ability of its members to utilize technologies such as cameras, smartphones, printers, and scanners while in the workplace,[210] an organization's ability to restrict the use of such technologies by its employees during their personal, non-work time and away from workplace facilities is limited. This causes challenges for information security, insofar as the same sensitive, work-related information that was captured, generated, stored, or transmitted by neuroprosthetic devices during working hours in the workplace may also be present in or recoverable from the devices when their human hosts are away from the workplace and engaging in purely private activities.

H. Account management

1. Automatic removal of temporary and emergency accounts

Implementing the automatic deletion of emergency accounts after a predetermined time[211] (rather than through manual action of a device's operator) may create a potential danger to the health and safety of a device's host, if an emergency account were being used to access the device in order to perform an urgent repair or provide some emergency medical service.

2. Automatic inactivity logouts

Implementing an automatic logout after a predetermined period of inactivity[212] should be done only after careful consideration, given the fact that a device's operator or host may expect and sometimes require instantaneous access to the device's functionality, and the delay caused by a need to log into an account would be unacceptable and potentially hazardous.

3. Disabling of accounts for high-risk users

Even if the operator or human host of a neuroprosthetic device has been clearly identified as a 'high-risk individual' who is likely to use the device for unauthorized purposes or to be targeted in whaling attacks, the decision of whether to disable the individual's account may raise serious legal and ethical questions, if disabling the account could impair (or even wholly terminate) the functioning of the device, thereby causing physical or psychological harm to the device's host or to others who would in some way be affected.[213]

[210] *NIST SP 800-53* (2013), p. F–214.

[211] *NIST SP 800-53* (2013), p. F–9.

[212] *NIST SP 800-53* (2013), p. F–9.

[213] Regarding the disabling of accounts for high-risk individuals, see *NIST SP 800-53* (2013), p. F–

4. Restrictions on privileges for non-organizational users

Fully blocking non-organizational users from exercising privileged access[214] to a neuroprosthetic system may not be possible if the system must possess mechanisms allowing privileged access (e.g., by medical personnel) in the case of a medical emergency affecting a device's human host.[215]

I. Security awareness training

Security awareness training[216] is important not only for the hosts or operators of neuroprosthetic devices but also for all individuals who live, work, or otherwise spend time in environments in which it is possible that other persons may possess neuroprosthetic systems that would allow them to compromise the individuals' information security.

J. Analyzing vulnerabilities in the deployed production context

1. Attack surface reviews

An organization may conduct attack surface reviews to identify physically or electronically exposed elements of an information system that increase its vulnerability to attacks; such attack surfaces include "any accessible areas where weaknesses or deficiencies in information systems (including the hardware, software, and firmware components) provide opportunities for adversaries to exploit vulnerabilities."[217] In the case of advanced neuroprostheses, attack surfaces may comprise not only the hardware and software components of a device itself but also anatomical structures, biological systems, and cognitive processes within a device's human host.

2. Penetration testing

The traditional conceptualization of penetration testing[218] as either black-, gray-, or white-box testing takes on new aspects in the case of advanced neuroprosthetics. In a sense, it may be impossible for any developer (or outside assessor acting on behalf of the developer) playing the role of an adversary to conduct full white-box testing, insofar as that would entail being given all available schematics, documentation, and information relating to the functioning of the system – and in the case of an advanced neuroprosthetic

---

10.

[214] *NIST SP 800-53* (2013), p. F–20.

[215] See Chapter Three of this text for an in-depth discussion of issues relating to emergency access to neuroprosthetic devices for medical personnel.

[216] *NIST SP 800-5* (2013), p. F–37.

[217] *NIST SP 800-53* (2013), p. F–168. See Chapter Two of this text for a discussion of vulnerabilities of neuroprosthetic devices. For vulnerabilities in IMDs generally, see Hansen & Hansen (2010).

[218] *NIST SP 800-53* (2013), p. F–168.

device, some such information and resources may be stored in the mind of the device's host in a way that cannot be conveyed to any other party; in such a situation, only a device's human host could (if sufficiently skilled) perform true white-box penetration testing.

3. Penetration testing by independent agents

Allowing penetration testing by independent agents or teams[219] may create special dangers, insofar as independent agents who lack full access to information about the nature of a neuroprosthetic device and its human host may inadvertently employ penetration technologies or techniques that are especially likely to cause harm to that host. On the other hand, independent agents are free from conflicts of interest that may arise with penetration testing conducted by an organization's internal personnel.

4. Red team exercises

The potential use of penetration testing to identify vulnerabilities or test the resistance of an advanced neuroprosthetic device in use within a human host to hostile cyberattacks, social engineering, espionage, and other efforts at compromising information security must be carefully considered, given the possibility that such testing[220] (whether or not it is successful in exploiting vulnerabilities) may cause physical or psychological harm to the device's host or others. Legal and ethical questions arise surrounding the extent to which penetration testing may be conducted on a neuroprosthetic device that has already been implanted in a human host; however, in some cases it may be impossible to accurately simulate the performance of an implantable device outside of the unique circumstances of its implantation within its particular host. Moreover, if vulnerabilities indeed exist, penetration testing may allow them to be discovered by the neuroprosthetic device's operator and addressed before they can be exploited by hostile outside parties who might intentionally exploit them to inflict maximum possible damage to the device's human host.

5. Penetration testing of physical facilities

In the case of advanced neuroprosthetic devices, it may or may not be permissible for a device's operator to conduct penetration testing that involves "unannounced attempts to bypass or circumvent security controls associated with physical access points,"[221] if such operations create a risk that physical or psychological harm may result to a device's human host or others.

---

[219] *NIST SP 800-53* (2013), p. F–62.

[220] Regarding red team exercises, see *NIST SP 800-53* (2013), p. F–62.

[221] *NIST SP 800-53* (2013), p. F–130.

6. Active testing of devices' response to known malicious code

A neuroprosthetic device's mechanisms for protecting the device and host-device system against malicious code can be tested "by introducing a known benign, non-spreading test case into the information system."[222] Great care should be taken and all legal, ethical, and practical implications considered before intentionally introducing such code into a neuroprosthetic device that is already integrated into the neural circuitry of a human host, as code that had previously been believed to be "benign" and "non-spreading" when studied in a laboratory setting may behave in unpredictable ways when exposed to or affected by the unique biological structures or activity of a particular human host.

SDLC stage 4: device operation within the host-device system and supersystem

The fourth stage in the system development life cycle includes the activities occurring after a neuroprosthetic device has been deployed in its production environment (comprising its host-device system and broader supersystem) and is undergoing continuous use in real-world operating conditions. The development or execution of security controls in this stage of the SDLC is typically performed by a device's operator and maintenance service provider(s) with the active or passive participation of its human host. Such controls are considered below.

A. Operations security (OPSEC)

1. Intentional misdirection to conceal information systems and their characteristics

An organization may utilize practices such as virtualization techniques, the intentional promulgation of believable but misleading information about the organization's systems or operations, the concealment of system components, and **deception nets** (including **honeynets** that intentionally utilize outdated or poorly configured software) in order to confuse adversaries and potentially lead them to undertake attacks that will be ineffective.[223] In the case of advanced neuroprosthetic devices, mechanisms designed for concealment and misdirection may need to be able to distinguish, for example, between an adversary who is attempting to break into and take control of a device for malicious purposes and emergency medical personnel who are attempting to 'break into' and take control of a device in order to save its host from some

---

[222] *NIST SP 800-53* (2013), p. F–218.

[223] *NIST SP 800-53* (2013), pp. F–205-06.

life-threatening medical danger. In the latter case, mechanisms for conceal-ment or misdirection purposefully added to a device or its software by their developers could potentially result in financial liability and legal and moral responsibility for the developers in the case of physical or psychological harm that is caused to the device's host or others as a result of emergency respond-ers being actively slowed or misdirected by such mechanisms.[224]

2. Concealment and randomization of communications

An adversary who is unable to gain access to the exact contents of com-munications may nonetheless obtain valuable intelligence by being able to observe phenomena such as the "frequency, periods, amount, and predicta-bility" of communications.[225] Mechanisms or practices that conceal or obscure such patterns can contribute to the information security of a neuroprosthetic device; however, the ability to randomize or conceal such communications may be limited by practical functional considerations such as the need to communicate effectively with the biological systems of a device's host and the fact that many forms of communication typically utilized by a human being – and thus a host-device system (such as speech, paralanguage, and gestures) – are effective precisely because they release information into an external en-vironment in a way that is not concealed or obscured.

3. Controlling physical access to devices outside of the organizational environment

Maintaining physical access control[226] is a challenge in the case of ad-vanced neuroprosthetic devices. The fact that a device is implanted within the body of a human host creates a practical, legal, and ethical obstacle that may prevent casual attempts by unauthorized parties to access the device: a neuroprosthetic unit that is implanted deep within a host's brain and pos-sesses no external physical access ports is more difficult to physically access than a computer sitting on a desktop in an exposed workplace environment.[227] On the other hand, the fact that a neuroprosthetic device is implanted in a human host who can conceivably take it anywhere in the world – and who could potentially be abducted and forcibly restrained or transported – in-creases the opportunity to gain physical access to the device for unauthorized

---

[224] See Chapter Three of this text for some proposed approaches to shielding or jamming tech-nologies that mask or conceal a neuroprosthetic device's existence but which can be disabled by emergency medical personnel when necessary.

[225] Regarding such threats and the approaches to communication concealment and randomiza-tion that can be employed to counteract them, see *NIST SP 800-53* (2013), p. F–194.

[226] *NIST SP 800-53* (2013), pp. F–128-29.

[227] See the device ontologies in Chapters One and Two of Gladden (2017) for ways in which the information security of a neuroprosthetic device can be affected by the device's physical struc-ture and its location within or in relation to its host's body.

parties that have sufficient means and motivation, especially if a device possesses visible and easily accessible external slots, ports, or other physical access points. This places greater demands on an organization's OPSEC personnel to protect such devices and their hosts.

B. Control of device connections

1. Protections against unauthorized physical connections

In the case of advanced neuroprosthetic devices, unauthorized physical connections with a device[228] (or its larger host-device system) might come not only through the connection of unauthorized external electronic devices to electronic components of the neuroprosthetic device but also through the presence of biological or biochemical agents and vectors (such as viruses, microorganisms, nootropic drugs, or other chemicals or substances) that can enter a host's organism and interface with his or her biological systems.[229]

2. Automatic termination of network connections

The automatic termination of a neuroprosthetic device's network connection after an arbitrary predetermined period of time could potentially result in physical or psychological harm to the device's host or user if the termination occurred during the midst of some critical activity.[230] Some naturally occurring biological cycles that are present within the biological systems and processes of a device's host (e.g., sleep cycles or cycles of neuronal firing) may provide opportunities for the safe deallocating and reallocating of address/port pairings, the disconnecting and reconnecting of network services, or other kinds of regular processes needed for maintaining a device's security and functionality.

C. Media protections

1. Controls on access to storage media

It may be impractical, unethical, and illegal for an organization – in its effort to control access to storage media[231] – to attempt to dictate, for example, that information system media remain within the organization's secured facility when the media are contained within neuroprosthetic devices implanted in the bodies of human hosts; it may not be possible to control the

---

[228] Regarding protections against unauthorized physical connections, see *NIST SP 800-53* (2013), p. F–191.

[229] Neuroprosthetic devices that include biological components may be especially liable to such attacks. For the possibility of neuroprosthetic devices involving biological components, see Merkel et al. (2007); Rutten et al. (2007); and Stieglitz (2007).

[230] Regarding automated network disconnection, see *NIST SP 800-53* (2013), p. F–194.

[231] *NIST SP 800-53* (2013), pp. F–119-21.

location of a storage medium without controlling (whether legally or unlaw-fully) the location of the human being in whom it is situated. If the infor-mation contained within a neuroprosthetic device is sufficiently valuable, an organization may not be able to assume that adversaries will not threaten, physically restrain, abduct, or harm the human host in whom the infor-mation's storage medium is housed in order to gain access to it.

2. Restrictions on media transport

It may be difficult or impossible to document or restrict the transporting of storage media[232] if they are contained in neuroprosthetic devices implanted in human hosts whose movements cannot legally or ethically be constrained or precisely tracked.

3. Implications of access for portable storage media

Some neuroprosthetic devices that possess an external port, media slot, or socket may allow data to be easily copied to or from portable storage media or devices, with significant implications for information security.[233]

D. Exfiltration and other output protections

1. Access controls for output mechanisms

Some neuroprosthetic devices not only include (or are connected to) tra-ditional output devices such as radio transmitters or monitors; they are also linked to 'output devices' such as the voice-box, facial muscles, hands, and other motor organs of their human host, which can be used to produce out-put in the form of speech, facial expressions, hand gestures, or typed or writ-ten communication.[234] Attempting to limit a host's use of such output systems may not be legally or ethically appropriate.[235]

2. Filtering of device output

In the case of advanced motor neuroprosthetics, filtering[236] may be used to (perhaps only temporarily) prevent the execution or expression of motor behavior that is identified as being anomalous and inconsistent with the kinds of motor behaviors expected from a device and its host-device system

---

[232] *NIST SP 800-53* (2013), p. F–121.

[233] See the device ontology in Chapter One of Gladden (2017) for a discussion of such components of neuroprosthetic devices. Regarding portable storage devices, see *NIST SP 800-53* (2013), p. F–33.

[234] See the device ontology in Chapter One of Gladden (2017) for a discussion of different output mechanisms for neuroprosthetic devices.

[235] Regarding access controls for output devices, see *NIST SP 800-53* (2013), pp. F–130-31.

[236] Regarding information output filtering, see *NIST SP 800-53* (2013), p. F–232.

in some given circumstances. This could potentially prevent a motor neuro-prosthesis from being hijacked by an adversary and used to perform an action that might disclose sensitive information or cause physical or psychological harm, embarrassment, or other negative impacts for the device's human host, operator, or others.[237] At the same time, care must be taken that such filters do not prevent a device's host or operator from expressing legally and ethically permissible motor actions that are fully intended by the host or operator simply because they are unusual and determined by the automated filter to be anomalous or suspicious.

3. Prevention of unauthorized exfiltration of information

The unauthorized **exfiltration** of information from a neuroprosthetic device or host-device system can potentially be detected and prevented through practices such as monitoring a device's communications to detect **beaconing** from within the device (e.g., directed at an external command-and-control server from which the compromised device is awaiting instructions), analyzing outgoing communications to detect **steganography**, and using traffic profile analysis to detect other anomalous communications that may potentially indicate exfiltration.[238] In the case of neuroprosthetic devices that control or support the cognitive processes or motor activity of their human host, care must be taken to ensure that mechanisms designed to prevent unauthorized exfiltration do not slow, block, or otherwise impede a host's communications and interaction with the external environment in a way that could result in physical or psychological harm to the host or others. In some circumstances, it may not be legally or ethically permissible to immediately block outgoing communications – even if an occurrence of ongoing unauthorized exfiltration has been confirmed – if impeding the outgoing communications could have sufficiently negative consequences for the survival or health of the device's host.

4. Mechanisms for the controlled release of information

It may be difficult or impossible to prevent the release of information[239] beyond the boundaries of a neuroprosthetically augmented information system if the mind of a neuroprosthetic device's human host is a part of that system, as the mind can express and convey information through speech, gestures, and other means that are not readily controlled.

---

[237] For the possibility of a neuroprosthetic limb being hacked by an adversary in order to manipulate its motor activity, see Denning et al. (2009).

[238] *NIST SP 800-53* (2013), p. F–190.

[239] *NIST SP 800-53* (2013), p. F–13.

E. Maintenance

1. Controls on the timing and location of maintenance activities

Conducting maintenance procedures on an advanced neuroprosthesis may require performing a surgical operation on the device's host, which would necessitate close coordination with the host and medical personnel. Even in cases when no surgical procedures are required, maintenance operations should be planned and scheduled in such a way that they do not cause undue interruption or impairment to a host's cognitive and physical capacities and, in particular, that they do not cause physical or psychological harm to the host or others. Conducting all maintenance within a secure facility may be desirable in order to ensure, for example, that automated maintenance instructions that are sent remotely to a neuroprosthetic device and which will result in a temporary device outage or change in the unit's functionality do not arrive when the device's host is engaged in performing a critical or potentially dangerous task.[240]

2. Control of maintenance equipment and software

Standard practices which prevent the unauthorized removal of system maintenance software or tools from a device and which restrict their use[241] may not be appropriate for neuroprosthetic devices that are a part of their host's organism; legal and ethical considerations may dictate that the host have full access to maintenance tools, including the ability to remove them. Emergency medical personnel treating the host may also need immediate unfettered access to some system maintenance tools that will allow them to affect or control the device's current operations, even if they are not provided full (or even partial) access to information stored within the device.[242]

3. Oversight of maintenance personnel

Efforts by a neuroprosthetic device's operator to limit who is allowed to conduct maintenance activities on the device – thereby restricting its host's ability to select his or her own maintenance personnel – may not be legal or ethical, given the device's status as an implantable device that may be considered an integral part of the host's body.[243]

Given the extremely large financial commitment that may be involved with acquiring authorized replacement parts and maintenance services for

---

[240] Regarding different possible service outage or maintenance schedules and their impact on a neuroprosthetic device's availability, see Chapter Three of this text. Regarding controls on the timing and location of maintenance procedures, see *NIST SP 800-53* (2013), p. F–112.

[241] *NIST SP 800-53* (2013), pp. F–113-14.

[242] See Chapter Three of this text for a discussion of different approaches to providing medical personnel with emergency access to a neuroprosthetic device.

[243] Regarding the control of maintenance personnel, see *NIST SP 800-53* (2013), p. F–116.

some kinds of neuroprosthetic devices, there may be strong financial incentives for the human hosts (or potentially operators) of such devices to seek out replacement components and services through less expensive unauthorized black- or gray-market channels offering pirated or counterfeit components and services that lack quality guarantees or warranties and are provided by individuals lacking formal training, licensing, or insurance. Such unauthorized channels may sometimes offer products that are more expensive than their authorized counterparts because they are free from standard security or DRM mechanisms or have been legally banned or offer services that cannot legally be provided or received.[244]

4. Predictive maintenance

Efforts to require the host of a neuroprosthetic device to submit to mandatory preventative device maintenance on the basis of predictive algorithms, a fixed calendar, or an *ad hoc* decision on the part of the device's operator may or may not be legal, if the maintenance may impact the host's cognitive or physical functioning and he or she does not wish to submit to it.[245]

5. Prevention of predictable failures

A best practice is to determine the mean time to failure (MTTF) for information system components not simply by relying on reported industry averages but by calculating the MTTF for components as they are used in particular installations by an organization.[246] Knowing the MTTF for components in use helps the organization to ensure that it has an adequate supply of replacement components on hand and is ready to repair or replace components when needed.

For some kinds of neuroprosthetic devices, the MTTF for individual components or a device as a whole may be influenced or determined by factors relating to the unique biological structures or processes of the device's individual human host. In such cases, it may be impossible to accurately estimate the MTTF for components in a particular device until the device has been put into operation and components have begun to fail.

---

[244] For the possibility of hosts modifying their own devices in unanticipated and potentially unwise and illicit ways, see Denning et al. (2010).

[245] For a discussion of predictive maintenance, see *NIST SP 800-53* (2013), p. F–118.

[246] Regarding predictable failure prevention, see *NIST SP 800-53* (2013), p. F–231. See Chapter Three of this text for a discussion of mean time to failure, mean time to repair, and availability for advanced neuroprosthetic devices.

F. Transmission of security alerts, advisories, and instructions

In the case of some kinds of sensory or cognitive neuroprostheses, it may be possible for an organization to deliver a security alert or directive[247] instantaneously and directly to the conscious awareness of a device's host through sensory input or augmented reality. However, if such methods are used by an organization as the primary or only way of delivering such alerts and directives, care must be taken to ensure that this delivery system cannot be blocked, disrupted, or manipulated by an adversary in order to facilitate a cyberattack on the device's host.

SDLC stage 5: device disconnection, removal, and disposal

The fifth stage in the system development life cycle involves a neuroprosthetic device's functional removal from its host-device system and broader supersystem; this may be accomplished through means such as remote disabling of the device or its core functionality, surgical extraction of the device, or the device's physical disassembly or destruction. The stage also includes a device's preparation for reuse or ultimate disposal after removal from its previous human host. The development or execution of security controls in this stage of the SDLC is typically performed by a device's operator or maintenance service provider(s), potentially with the active or passive participation of its human host. Such controls are considered below.

A. Procedures for information retention

Individuals and organizations may be required to retain some information that is received, generated, stored, or transmitted by neuroprosthetic devices for legal, ethical, or practical reasons.[248] Note that some kinds of neuroprosthetic devices that mimic or interface with the natural biological memory systems of the human brain may store information in a way that is subject to significant compression, distortion, and degradation over time.[249] While storing information in our natural biological memory systems has, throughout human history, often been the best or only way of storing such information, the use of neuroprostheses that demonstrate such functional limitations may not be legally, ethically, or operationally advisable in cases when more effective and reliable storage mechanisms are available.

---

[247] Regarding security alerts, advisories, and directives, see *NIST SP 800-53* (2013), p. F–224.

[248] For a discussion of information retention policies and procedures, see *NIST SP 800-53* (2013), p. F–230.

[249] Regarding questions surrounding the nature and quality of long-term memory storage in the human brain, see Dudai (2004).

B. Sanitization of media prior to reuse or disposal

Neuroprosthetic devices or component storage units removed from a human host may contain confidential information about the host's biological processes and sensory experiences that must be cleared, purged, or destroyed before the device can be released for reuse or disposal.[250] Destruction of a storage medium may not be necessary if it can be guaranteed that the information cannot be retrieved from the medium or otherwise reconstructed. In the case of storage media contained within neuroprosthetic devices implanted within a human host, it may be impractical, illegal, and unethical to attempt to erase, purge, or destroy a storage medium without (or potentially even with) the host's permission.

Conclusion

In this chapter, we have reviewed a number of standard preventive security controls for information systems and discussed the implications of applying such controls to neuroprosthetic devices and the larger information systems in which they participate, using the lens of a five-stage system development life cycle as a conceptual framework. In the following chapters, a similar analysis of detective and corrective or compensating controls will be undertaken.

---

[250] Regarding media sanitization, see *NIST SP 800-53* (2013), pp. F–122-23.

Chapter Seven

Detective Security Controls for
Neuroprosthetic Devices and Information Systems

Abstract. This chapter explores the way in which standard detective security controls (such as those described in *NIST Special Publication 800-53*) become more important, less relevant, or significantly altered in nature when applied to ensuring the information security of advanced neuroprosthetic devices and host-device systems. Controls are addressed using an SDLC framework whose stages are (1) supersystem planning; (2) device design and manufacture; (3) device deployment; (4) device operation; and (5) device disconnection, removal, and disposal.

Detective controls considered include those relating to the establishment of an integrated InfoSec security analysis team; use of all-source intelligence regarding component suppliers; integrity indicators; designing the capacity to detect medical emergencies; integrated situational awareness; establishment of account usage baselines; general monitoring and scanning; auditing of events; threat and incident detection; and proactive detection and analysis methods.

Introduction

In this chapter, we explore a range of standard detective security controls for information systems and identify unique complications that arise from the perspective of information security, biomedical engineering, organizational management, and ethics when such controls are applied to neuroprosthetic devices and larger information systems that include neuroprosthetic components. The text applies such security controls without providing a detailed explanation of their basic nature; it thus assumes that the reader possesses at least a general familiarity with security controls. Readers who are not yet acquainted with such controls may wish to consult a comprehensive catalog such as that found in *NIST Special Publication 800-53, Revision 4*, or *ISO/IEC 27001:2013*.[1]

---

[1] See *NIST Special Publication 800-53, Revision 4: Security and Privacy Controls for Federal Information Systems and Organizations* (2013) and *ISO/IEC 27001:2013, Information technology – Security techniques – Information security management systems – Requirements* (2013).

Approaches to categorizing security controls

Some researchers classify controls as either **administrative** (i.e., comprising organizational policies and procedures), **physical** (e.g., created by physical barriers, security guards, or the physical isolation of a computer from any network connections), or **logical** (i.e., enforced through software or other computerized decision-making).[2] Other sources have historically categorized controls as either **management**, **operational**, or **technical** controls. As noted in the previous chapter, in this volume we follow the lead of texts such as *NIST SP 800-53*,[3] which has removed from its security control catalog the explicit categorization of such measures as management, operational, or technical controls, due to the fact that many controls reflect aspects of more than one category, and it would be arbitrary to identify them with just a single category. We instead utilize a classification of such measures as **preventive**, **detective**, or **corrective and compensating** controls. The previous chapter considered the first type of control; this chapter investigates the second type; and the subsequent chapter will explore the third and final type.

Role of security controls in the system development life cycle

The detective controls discussed here are organized according to the stage within the process of developing and deploying neuroprosthetic technologies when attention to a particular control becomes most relevant. These phases are reflected in a system development life cycle (SDLC) whose five stages are (1) supersystem planning; (2) device design and manufacture; (3) device deployment in the host-device system and broader supersystem; (4) device operation within the host-device system and supersystem; and (5) device disconnection, removal, and disposal.[4] Many controls relate to more than one stage of the process: for example, the decision to develop a particular control and the formulation of its basic purpose may be developed in one stage, while the details of the control are designed in a later stage and the control's mechanisms are implemented in yet a further stage. Here we have attempted to locate a control in the SDLC stage in which decisions or actions are undertaken that have the greatest impact on the success or failure of the given control. This stage-by-stage discussion of detective controls begins below.

---

[2] Rao & Nayak, *The InfoSec Handbook* (2014), pp. 66-69.

[3] See *NIST SP 800-53* (2013).

[4] Various approaches to defining the stages of an SDLC for an information system involving neuroprosthetic components are reviewed in Gladden, "Managing the Ethical Dimensions of Brain-Computer Interfaces in eHealth: An SDLC-based Approach" (2016).

SDLC stage 1: supersystem planning

The first stage in the system development life cycle involves high-level planning of an implantable neuroprosthetic device's basic capacities and functional role, its relationship to its human host (with whom it creates a biocybernetic host-device system), and its role within the larger 'supersystem' that comprises the organizational setting and broader environment within which the device and its host operate. The development of security controls in this stage of the SDLC typically involves a neuroprosthetically augmented information system's designer, manufacturer, and eventual institutional operator.

A. Establishment of an integrated InfoSec security analysis team

While many protective controls are relevant in the planning stage of the SDLC, only one detective control sees its critical moment occur during that stage: the establishment of an integrated InfoSec security analysis team that can detect and analyze vulnerabilities, threats, and incidents that occur throughout the remaining stages of the SDLC. In the case of advanced neuroprosthetic devices, an integrated information security analysis team may need to incorporate not only typical members such as "forensic/malicious code analysts, tool developers, and real-time operations personnel"[5] but potentially also biomedical engineers, biologists, neuroscientists, psychologists, biocyberneticists, and implantation surgeons.[6]

SDLC stage 2: device design and manufacture

The second stage in the system development life cycle includes the design and manufacture of a neuroprosthetic device and other hardware and software that form part of any larger information system to which the device belongs. The development of security controls in this stage of the SDLC is typically carried out by a device's designer and manufacturer, potentially with instructions or other input from the system's eventual operator. Such controls are considered below.

A. Use of all-source intelligence regarding component suppliers

The potential widespread use of advanced neuroprostheses by the public (including by the employees and customers of an organization's suppliers) may provide organizations with a new element to incorporate into all-source

---

[5] *NIST SP 800-53* (2013), p. F–110.

[6] See Chapter Three of this text for a discussion of the growing interconnection of information security with fields such as neuroscience and biomedical engineering, especially in the context of advanced neuroprosthetic devices.

intelligence analysis: namely, the thoughts, memories, perceptions, plans, and emotions of individuals associated with current or potential suppliers that are publically shared by these persons through neuroprosthetically enabled social networks. This would represent a potentially deeper and more sophisticated source of information and analysis than can be obtained, for example, from the analysis of contemporary social media posts.[7]

B. Design of integrity indicators

1. Integrity checks for firmware and software

Checking the integrity of a device's operating system or applications[8] may be difficult or impossible in the case of some neuroprosthetic devices that utilize physical neural networks (and do not execute 'programs' as conventionally understood) or which are passive devices that are directly controlled by their host's cognitive processes, which effectively provide the 'operating system' for the device.[9] In the case of neuroprosthetic devices that utilize biological components for storing information and performing activities, it may be impossible to require the same level of integrity as that expected with electronic computers, insofar as the biological components may be undergoing gradual but continuous change through the birth, growth, mutation, and death of individual cells.

2. Tamper-detection mechanisms

Tamper-detection seals and anti-tamper coatings[10] may be utilized to prevent unauthorized access to a neuroprosthetic device's internal components or to ensure that if such components *have* been accessed by an unauthorized party, evidence of that unauthorized access will be visible to the next authorized party who conducts routine maintenance operations on or provides other service for the device.[11]

C. Designing the capacity to detect medical emergencies

A neuroprosthetically augmented information system may not only be able to detect errors and incidents relating to the electronic portion of the system but may also be able to directly or indirectly detect medical incidents

---

[7] Regarding all-source intelligence and component system suppliers, see *NIST SP 800-53* (2013), p. F–171.

[8] *NIST SP 800-53* (2013), p. F–225.

[9] See Chapter One of this text for a discussion of passive neuroprosthetic devices and Chapter Two for a discussion of integrity as an information security goal and attribute.

[10] *NIST SP 800-53* (2013), p. F–129.

[11] The systems described in Chapter Three of this text for providing audible – rather than visible – alerts to a device's host when attempts are made to wirelessly access the device constitute another kind of anti-tampering mechanism.

and other biological problems affecting to the human host of an implanted neuroprosthesis. For example, Rasmussen et al. have proposed a model of emergency access control for implantable medical devices that relies on ultrasound technology to verify the physical proximity of an external system attempting to gain access to an IMD. Normally the IMD would require an external system to possess a shared cryptographic key before granting the external system access to the IMD; however if the IMD detects that its host is undergoing a medical emergency, it shifts into an 'emergency mode' in which any external system is allowed to access the IMD, as long as it is within a certain predefined distance, as measured by the time required for ultrasound communications to travel between the IMD and external system.[12]

SDLC stage 3: device deployment in the host-device system and broader supersystem

The third stage in the system development life cycle includes the activities surrounding deployment of a neuroprosthesis in its human host (with whom it forms a biocybernetic host-device system) and the surrounding organizational environment or supersystem. The development or implementation of security controls in this stage of the SDLC is typically performed by a device's operator with the active or passive participation of its human host. Such controls are considered below.

A. Fostering of integrated situational awareness

Organizations integrate information obtained "from a combination of physical, cyber, and supply chain monitoring activities"[13] in order to better detect cyberattacks, which may be multifaceted operations that have both physical and virtual components and which target both an employer's operation of a neuroprosthesis and the suppliers who designed and produced the device's hardware and software components. For individual human hosts who operate neuroprostheses that they have purchased or leased as consumer electronics devices, developing such integrated situational awareness relating to their devices can be difficult. For such an individual, the physical monitoring of his or her device may be relatively easy, insofar as the device is always physically present with the host, and in order for unauthorized parties to physically access the device they may need to physically access or manipulate the host's biological body. Awareness of cyberattacks or other unauthorized

---

[12] See Rasmussen et al., "Proximity-based access control for implantable medical devices" (2009). Regarding the possibility of IMDs being able to detect a medical emergency that is being experienced by their human host, see Denning et al., "Patients, pacemakers, and implantable defibrillators: Human values and security for wireless implantable medical devices" (2010), pp. 921-22.
[13] *NIST SP 800-53* (2013), p. F-222.

electronic access may be more difficult for the host to achieve and may depend on a combination of effective security controls built into the device, its OS, and its applications, as well as personal knowledge of and commitment to InfoSec best practices on the part of the device's host. Full supply chain monitoring may be difficult or impossible for the host to carry out: although he or she may know the identity of the organizations that were responsible for assembling and distributing the finished physical device and its operating system and installed applications, it may be difficult for the host to determine who designed and manufactured individual components within the device or who may have served as a subcontractor writing and testing outsourced portions of the OS and applications on behalf of the primary developer. In the case of open-source software, it may be more difficult to know the true identity of the parties responsible for providing particular elements of code, although it may simultaneously be easier to scrutinize the content of the code itself.

B. Establishing baselines to detect atypical account usage

For some kinds of neuroprosthetic devices, it may be difficult to establish clear baselines and a definition of what constitutes 'typical' and 'atypical' usage,[14] just as it is difficult to clearly define what constitutes 'typical' thoughts, emotions, beliefs, volitions, or use of the imagination. This challenge may be exacerbated when a neuroprosthetic device is used to allow a user to interface with and experience some virtual environment that is, in a sense, already 'atypical' and likely to generate new kinds of activities and experiences.[15]

C. Activation of monitoring and scanning systems

1. Continuous monitoring, guards, and alarms

In the case of implantable neuroprosthetic devices, it may be possible to use one device as a 'guard' that monitors physical access to other devices implanted within the same human host.[16]

---

[14] *NIST SP 800-53* (2013), p. F–10.

[15] For some examples of neuroprosthetic devices that provide their hosts and users, e.g., with the sensorimotor experience of a new body that is (perhaps even radically) 'nonhuman,' see Gladden, "Cybershells, Shapeshifting, and Neuroprosthetics: Video Games as Tools for Posthuman 'Body Schema (Re)Engineering'" (2015).

[16] Regarding continuous guards, alarms, and monitoring, see *NIST SP 800-53*, (2013), p. F–129. See the related discussion in Chapter Three of this text of proposed schemes for emergency access to IMDs that utilize external cloaking devices or gateway devices to mediate, limit, or control access to an implanted device.

2. Specialized devices for information system monitoring

Significant legal, ethical, and practical questions arise regarding an organization's deployment of monitoring devices to observe and scrutinize an organizational information system.[17] Although such monitoring may have the legitimate purpose of detecting or dissuading attacks, it may sometimes also gather personal information on the activities, health, and other characteristics of organizational members or outside parties in ways that is legally and ethically impermissible. External systems (e.g., medical imaging or diagnostic equipment) may be used to monitor the activities of a neuroprosthetic device or host-device system; alternatively, a neuroprosthetic device may itself be used by an organization as a monitoring device to scrutinize the activity of other conventional information systems belonging to the organization (e.g., servers or desktop computers). Other implantable devices that are located within the same organism as a neuroprosthetic device may be used to monitor the activities of the device and its host-device system.[18]

3. Vulnerability scans

In some circumstances, even the mere act of scanning an implanted neuroprosthetic device to identify vulnerabilities[19] could be considered an invasive medical procedure and an infringement on the privacy of the human host in whom the device is implanted. In other circumstances, it might potentially be considered medical malpractice for an organization not to utilize all available means in probing neuroprosthetic devices implanted in its personnel to identify device vulnerabilities and the nature and extent of discoverable information within the devices and their connected systems that is potentially available to unauthorized parties.

4. Video surveillance

In the case of some neuroprosthetic devices such as artificial eyes, a device itself may be able to provide video surveillance[20] to its operator that records whether anyone has gained physical access to the device – with the caveat that if the device's security has already been compromised through some

---

[17] *NIST SP 800-53* (2013), pp. F–219-20.

[18] For a discussion of ethical and legal aspects relating to such issues, see Kosta & Bowman, "Implanting Implications: Data Protection Challenges Arising from the Use of Human ICT Implants" (2012); McGee, "Bioelectronics and Implanted Devices" (2008); Mak, "Ethical Values for E-Society: Information, Security and Privacy" (2010); McGrath & Scanaill, "Regulations and Standards: Considerations for Sensor Technologies" (2013); Shoniregun et al., "Introduction to E-Healthcare Information Security" (2010); and Brey, "Ethical Aspects of Information Security and Privacy" (2007).

[19] *NIST SP 800-53* (2013), p. F–153.

[20] *NIST SP 800-53* (2013), p. F–132.

other means (e.g., if it has been hacked through use of software that had been installed on the device through a wireless connection), it may not be possible to trust the accuracy or integrity of any video stream being provided by the device, as that imagery could be fabricated or altered.[21]

An artificial eye may also be able to provide video surveillance that will allow its operators to determine whether any parties have acquired physical access to other neuroprosthetic devices implanted in the same host or, potentially, in other persons who are within the artificial eye's field of vision.

5. Systematic intrusion detection mechanisms

Intrusions *into* neuroprosthetic devices may be detected by standard tools that monitor the electronic components and systems of a device; they may also potentially be detected as alterations in a device's functioning by the human host with whose neural circuitry the device is integrated. Intrusions into conventional information systems committed *using* neuroprosthetic devices may – depending on the nature of the intrusion – be detected by traditional intrusion-detection mechanisms,[22] be detected by specialized detection mechanisms designed specifically to recognize the presence and activity of neuroprosthetic devices, or be difficult to detect by any means.[23]

6. Surveillance equipment for intrusion detection

Multiple neuroprosthetic devices may be able to create a body area network (BAN) or body sensory network (BSN) in which devices conduct mutual surveillance, monitor one another's status, and identify physical intrusions into their host's body or bodily systems.[24]

7. Technical surveillance countermeasures surveys

Technical surveillance countermeasures surveys are conducted in order "to detect the presence of technical surveillance devices/hazards and to identify technical security weaknesses that could aid in the conduct of technical penetrations of surveyed facilities."[25] They are generally performed using a combination of intensive electronic testing, visual observation, and physical

---

[21] For the possibility that a neuroprosthesis designed to receive raw data from the environment might have that data replaced with other data transmitted from some external information system, see Koops & Leenes, "Cheating with Implants: Implications of the Hidden Information Advantage of Bionic Ears and Eyes" (2012). Regarding the possibility that neuroprostheses could be used to provide false data or information to their hosts or users, see also McGee (2008), p. 221.

[22] Regarding system-wide intrusion detection systems, see *NIST SP 800-53* (2013), p. F–220.

[23] See Chapter Three of this text for a discussion of neuroprosthetic devices as potential tools for use in launching cyberattacks or other kinds of attacks.

[24] Regarding intrusion alarms and surveillance equipment, see *NIST SP 800-53* (2013), p. F–131.

[25] *NIST SP 800-53* (2013), p. F–155.

examination of information systems, the facilities in which they are housed, and the surrounding environment.[26]

In the case of advanced neuroprostheses utilized by an organization, it may not always be legally, ethically, or practically feasible to conduct countermeasures surveys in all of the venues in which a neuroprosthetic device may operate (e.g., within the home of its human host), even with the advance consent of the host. Surveillance countermeasures surveys must also be conducted in a way that does not create a danger of physical or psychological harm for the host of a neuroprosthetic device or for others. Finally, it should be noted that in some cases the efficacy of surveillance countermeasures surveys that are planned and conducted in conjunction with the human host of a neuroprosthetic device may be compromised by the fact that one form of implementing a 'surveillance device' by adversaries would involve hacking a host's existing sensory organs, memory systems, or other cognitive processes in order to gain access to data gathered by or stored in a host's existing neuroprosthetic device.[27] In such a case, the adversary utilizing an existing neuroprosthesis as a surveillance device may – through the device – gain advance notice of planned surveillance countermeasures surveys and be able to evade them through appropriate planning. Moreover, some kinds of technical surveillance countermeasures surveys may be able to detect the presence of a surveillance device that should not have been present at all, but may have more difficulty detecting the fact that a human host's neuroprosthesis that was known to and whose presence was authorized by the organization had been hijacked or otherwise compromised by an adversary and was being (either temporarily, periodically, or permanently) employed by an unauthorized party as a surveillance device. It may be similarly difficult to detect situations in which an employee whose implanted neuroprosthesis is known to (and perhaps even provided by) his or her employer is being utilized by the employee in an unauthorized way as a surveillance device, particularly if the patterns of device activity reflected in such unauthorized uses are generally consistent with those seen when the device is used for authorized purposes.

8. Methods for detecting indicators of compromise

Organizations may use automated or manual procedures for searching for **indicators of compromise** (IOCs), which are detectable traces created or left within an information system that may indicate that the system has been compromised; such IOCs may include new registry key values or records of

---

[26] *NIST SP 800-53* (2013), p. F–155.

[27] See Chapter Three of this text for a discussion of the possibility of adversaries accessing another individual's neuroprosthetic device in order to create a surveillance instrument.

network traffic between the system and known command-and-control serv-ers.[28] In the case of advanced neuroprosthetic devices, IOCs may potentially take radically new and different forms, such as the presence of unexplained or corrupted memories or memory fragments within the mind of a device's host, the host's display of unusual sensory, motor, emotional, or personality-related behaviors, or the presence of particular hormones, other chemicals, or other objects within the host's bloodstream or body.[29]

SDLC stage 4: device operation within the host-device system and supersystem

The fourth stage in the system development life cycle involves the activi-ties occurring after a neuroprosthetic device has been deployed in its produc-tion environment (comprising its host-device system and broader supersys-tem) and is undergoing continuous use in real-world operating conditions. The development or execution of security controls in this stage of the SDLC is typically carried out by a device's operator and maintenance service pro-vider(s) with the active or passive participation of its human host. Such con-trols are considered below.

A. Ongoing general monitoring

1. Device monitoring and tracking

Monitoring and tracking the location[30] of an implanted neuroprosthetic device raises complex legal and ethical questions, insofar as this necessarily entails monitoring and tracking the location of the human host in whom it is implanted.

2. Incident monitoring

Conducting incident monitoring[31] to track and document security inci-dents may be difficult in the case of neuroprosthetic devices (such as those comprising biological components or nanorobotic swarms) that may lack a centralized mechanism capable of detecting and recording incidents affect-ing a device's components.[32]

---

[28] *NIST SP 800-53* (2013), p. F–223.

[29] For the possibility that an attack on a neuroprosthetic device or its host-device system might produce long-term changes in the neural activity or structures of the device's host, see Denning et al., "Neurosecurity: Security and Privacy for Neural Devices" (2009).

[30] *NIST SP 800-53* (2013), p. F–138.

[31] *NIST SP 800-53* (2013), p. F–107.

[32] See the discussion of passive neuroprosthetic devices in Chapter One of this text for examples of such devices.

3. Maintenance and scrutiny of visitor access records

It may be appropriate for the organization operating a neuroprosthetic device to record the identities and other details of individuals who gain direct physical access to the device itself; however, it may be legally or ethical questionable for the organization to record and archive details regarding the circumstances and identities of all individuals who gain physical access[33] to the device's host more generally (e.g., through face-to-face meetings or in other ways).

4. Collection and correlation of monitoring information

Given the resources needed for receiving, processing, and transmitting all of the monitoring data from a wide array of sources that is to be correlated,[34] such correlation may often be handled best by external systems that are managed by the operators of a neuroprosthetic device. The use of an external system that is housed within a conventional information systems facility avoids the severe limitations on memory storage, processing capacity, and communications bandwidth that affect many implantable neuroprosthetic devices due to their size, power, and operational constraints. Particular insights, conclusions, or instructions that result from the correlation of monitoring information in an external system can then be conveyed to a device or to its host or operator for use as appropriate.

C. Auditing of events

1. Specification of events to be audited

In the case of some neuroprostheses (such as those that utilize a physical neural network or are passive devices controlled by a host's neural circuitry[35]) it may be difficult to specify particular kinds of auditable events.[36]

2. Designing a storage system for audit data

Implantable neuroprosthetic devices may possess limited onboard capacity to store audit records generated by a device.[37] Offloading audit data to external systems for permanent storage may not be possible, for example, if a neuroprosthetic device includes a physical neural network with billions of neurons whose individual real-time actions have been designated as audit

---

[33] *NIST SP 800-53* (2013), p. F–132.

[34] *NIST SP 800-53* (2013), p. F–222.

[35] See the device ontology in Chapter One of Gladden, *Neuroprosthetic Supersystems Architecture* (2017), for a discussion of neuroprosthetic devices that utilize a physical neural network and Chapter One of this volume for a discussion of passive neuroprosthetic devices.

[36] *NIST SP 800-53* (2013), pp. F–41-42.

[37] Regarding audit storage capacity, see *NIST SP 800-53* (2013), p. F–43.

events but which cannot be recorded and transmitted to external systems using current or foreseeable technologies.

3. Sensitivity of audit data

Monitoring the ways in which its operator or host utilizes a neuroprosthetic device raises legal and ethical questions insofar as such monitoring may capture and record personal medical data, the contents of cognitive processes, or other sensitive data about the status and actions of the device's human host.[38]

4. Protections for audit data

The use of hardware-enforced write-once media, cryptographic protection, read-only access, and data backup on separate physical systems is beneficial for ensuring information security for a device's audit information.[39] However, the use of such technologies and techniques may or may not be possible for a neuroprosthesis. For example, an implanted device may have no means of backing up audit data to external systems.[40]

5. Chain of custody of audit data (non-repudiation)

The chain of custody of audit information[41] may be difficult or impossible to maintain for a neuroprosthetic device that stores audit information within itself (without the possibility of backup to external systems) and which is not stored permanently within a single secured facility belonging to the operator but rather implanted in a human being who brings the device into environments and situations in which unauthorized attempts to access the device may easily be made.

D. Threat and incident detection

1. Inspection of devices and components after deployment

The inspection of a neuroprosthetic device after its implantation may require the express consent of the device's host.[42] There is a danger that if the

---

[38] Regarding such legal and ethical issues, see, e.g., Hildebrandt & Anrig, "Ethical Implications of ICT Implants" (2012); Kosta & Bowman (2012); McGrath & Scanaill (2013); and Shoniregun et al. (2010).

[39] *NIST SP 800-53* (2013), p. F–49.

[40] See Chapter Four of this text for the distinction between information stored by neuroprosthetic devices in the form of engrams versus exograms; information stored using exograms is typically easier to back up to external systems.

[41] *NIST SP 800-53* (2013), p. F–50.

[42] Regarding the inspection of information systems, devices, and components after their deployment, see *NIST SP 800-53* (2013), p. F–180.

information security of an already-implanted device has indeed been compromised by an adversary – and the device is able to influence or exercise control over relevant biological or cognitive processes – then the human host may decline to express consent to an inspection not because the host has decided to reject the inspection through an act of his or her autonomous agency and volition but because the adversary has utilized control over the compromised device in a way that blocks the host from expressing agreement to an inspection.[43]

2. Security function verification during transitional states

In addition to regular ongoing security function verification, it is important to conduct special verification when a system is undergoing transitional states such as being powered on, rebooted, or shut down.[44] For some kinds of neuroprosthetic devices, key transitional states may also relate to the sensory, cognitive, and motor processes displayed by a device's human host, such as entering or leaving sleep, opening or closing eyelids, or initiating gross motor movements.

3. Non-signature-based detection of malicious code

Heuristic analysis and other approaches or mechanisms can be used to identify malicious code that is polymorphic or metamorphic and which cannot be identified by antivirus software that searches for particular known signatures.[45] In the case of some kinds of neuroprosthetic devices (e.g., those that utilize certain types of physical neural networks and do not execute programs as traditionally understood), malicious 'code' may not take the form of discrete strings of digital information that can be analyzed to detect particular signatures but may instead take the form of sense data or other forms of environmental phenomena that can affect a neuroprosthetic device or host-device system. In such situations, the use of heuristic analysis and probabilistic methods to identify potentially malicious input may be necessary.[46]

---

[43] See Chapter Three of this text for the related possibility of a neuroprosthetic device that unintentionally traps the mind of its human host within a 'zombie-like' host-device system in which the host is unable to express his or her thoughts or volitions using motor activity.

[44] *NIST SP 800-53* (2013), pp. F–224-25.

[45] *NIST SP 800-53* (2013), p. F–218.

[46] For some such neuroprosthetic systems, the process of detecting a malicious vector may be less like the discrete process of detecting a traditional computer virus (e.g., a kind of binary data file) and more like the ambiguous everyday challenge of identifying a potentially malicious *person*, a potentially damaging social relationship, or a potentially harmful sensory experience.

4. Analysis of malicious code to ascertain effects

In the case of neuroprosthetic devices that utilize biological components or materials, 'malicious code' may potentially take the form of genetic sequences delivered through the use of biological (and not computer) viruses, microorganisms, or other biological vectors.[47] A sophisticated attack on such a neuroprosthetic device might combine both the use of a computer virus or worm that infects and compromises electronic portions of the device and a biological vector that infects and compromises the biological portions of the device. In the case of a neuroprosthesis that controls, supports, or executes the production of hormones, cells, or other biochemical products within its host's body, a computer worm or virus or attack that compromises the electronic portion of the device could conceivably be used to generate biochemical agents that would in turn infect and compromise biological components of the neuroprosthetic device or of the host's natural organism. Conversely, a biological vector or biochemical agent that infects biological components of the host's organism or of a neuroprosthetic device could conceivably be used to introduce malware into the device's electronic components, if the device's electronic components receive and process information or other input from biological or biochemical components or systems within the host's organism or the neuroprosthetic device.

5. Detection of unauthorized commands

Information systems are often designed to detect unauthorized operating system commands at the level of the kernel application programming interface, block the execution of such commands, and issue an alert.[48] Such mechanisms are important not just for preventing certain kinds of attacks that are purposefully launched against information systems by adversaries but also for preventing unauthorized commands that may be an unintentional result of some hardware or software failure, quantum-level metastability, or other phenomenon. Guarding against the execution of unauthorized commands is especially important in the case of neuroprosthetic devices with critical health impacts for their human host.

6. Detection of communication or possession of unsanctioned information

Controls may be used, for example, to detect proprietary or classified information whose possession would be unlawful and to block it from being transferred into a host's memory by his or her neuroprosthetic device.[49]

---

[47] Regarding malicious code analysis, see *NIST SP 800-53* (2013), p. F–219.

[48] *NIST SP 800-53* (2013), p. F–218.

[49] For controls relating to unsanctioned information, see *NIST SP 800-53* (2013), p. F–17.

7. Identification and analysis of covert channels

An organization typically identifies ways in which devices or systems emit transmissions, material objects, or other phenomena that could be used as covert channels for communication; determines the maximum bandwidth available through such covert channels for potential unauthorized communications; and attempts to reduce the bandwidth available for covert channels, insofar as this is feasible given the organization's functional needs and operational priorities.[50] Some kinds of neuroprosthetic devices may not only generate phenomena such as wireless transmissions, magnetic fields, electrical charges, heat, biological or biochemical substances and materials, and other objects or phenomena that can be directly observed by external parties; the devices may also stimulate the body of their human host in a way that causes it to produce physical reactions or behaviors that can be observed by persons or sensors in the external environment and which can potentially serve as covert channels for the communication of information. It may not be legally, ethically, or practically possible to eliminate or minimize the bandwidth of all such channels.

8. Detection of anomalous communications traffic

In the case of general-purpose organizational information systems, anomalous communications traffic may include "large file transfers, long-time persistent connections, unusual protocols and ports in use, and attempted communications with suspected malicious external addresses."[51]

In the case of neuroprosthetic devices, anomalous traffic at the external boundaries of a device may result from unusually intense or numerous environmental stimuli impacting a sensory neuroprosthetic (perhaps reaching the level of sensory overload), unusually intense or complex motor instructions being sent to motor organs (e.g., when attempting to speak, play a musical instrument, or engage in sports activities), or unusually intense, rich, or complex cognitive activities (e.g., caused by or reflected in heightened emotion, a state of dreaming or hallucination, acts of mental calculation, the imagining of new ideas, or the retrieval of distant memories). Note that activities seen as producing 'anomalous' communications traffic patterns when viewed in relation to a neuroprosthetic device may appear quite commonplace from the perspective of the device's human host.

---

[50] *NIST SP 800-53* (2013), pp. F–206-07.
[51] *NIST SP 800-53* (2013), p. F–221.

9. Detection of wireless intrusions

Organizations may use a wireless intrusion detection system to scan for and detect both the connection of unauthorized wireless devices to the organization's own wireless access points as well as the presence of unauthorized wireless access points within organizational facilities.[52] Such wireless intrusion detection systems may be used, for example, to identify visitors to organizational facilities who are using nonvisible neuroprosthetic devices to make unauthorized connections to organizational information systems or to identify miniaturized neuroprosthetic devices that may have been implanted in organizational personnel without their knowledge and which are attempting to wirelessly contact command-and-control servers for instructions or to transmit gathered intelligence.[53]

10. Detection of extrusion and exfiltration attempts

The analysis of traffic to detect and prevent covert exfiltration[54] may be desirable and necessary even in the case of neuroprosthetic devices whose actions are clearly visible to their human host and which are theoretically engineered to transmit information or perform actions only in accordance with the volition of their host. For example, an advanced neuroprosthetic arm may be controlled by motor impulses originating in its host's brain, and its movements are visible to the eyes of its human host. However, a computer virus or adversary's cyberattack that is able to compromise the device and alter the motor instructions received by the device from its host's brain (or fabricate nonexistent motor instructions) may be able to cause the neuroprosthetic arm to move in minute ways or with subtly altered patterns that are undetectable to the host's natural biological eyes and visual perception but which can be detected and interpreted by external adversarial systems and used to exfiltrate information from the device or its host-device system.[55] In a sense, such a case would represent a form of **cognitive steganography** or **motor steganography**.

---

[52] *NIST SP 800-53* (2013), p. F–222.

[53] See Chapter Three of this text for a discussion of the reliance on wireless communication demonstrated by many kinds of neuroprosthetic devices.

[54] *NIST SP 800-53* (2013), p. F–222.

[55] Regarding the possibility that a neuroprosthetic limb could be hacked or otherwise compromised by an adversary and that its behavior could be remotely controlled or manipulated, see Denning et al. (2009).

E. Proactive detection and analysis methods

1. Use of devices as active honeyclients

It is theoretically possible to use neuroprosthetic devices (or their constituent components, subsystems, or subnetworks) as honeyclients[56] that proactively explore the Internet in search of malicious code – either code designed to infect and harm a device itself or (e.g., if the neuroprosthetic device is being used by cyberwarfare personnel within a military organization) designed to infect other kinds of systems that the operator of the neuroprosthetic device has an interest in protecting. However, the legal and ethical implications of such practices must be carefully considered, especially if they create an increased risk that the human host or operator of a neuroprosthetic device may experience physical or psychological damage as a result of the device's intentional encounter with malicious code.

2. Use of devices as passive honeypots

It may or may not be feasible to provide a neuroprosthetic device itself with the components, subsystems, or subnetworks needed to create a honeypot[57] that can either simply serve as a decoy that lures the attention of adversaries away from the device's actual core systems or which potentially allows adversaries' attacks to be observed and analyzed without creating a danger for the device's core systems. In many cases, it may not be an effective use of the limited resources that can be included in a small implantable device to create a honeypot within the device itself. Unique legal and ethical issues (including those of liability for possible damages to the host) may also arise through creating within the body of a human host such decoy systems that can attract attacks.

In some cases (e.g., those of neuroprosthetic devices that are composed largely or entirely of biological material, do not have significant mechanisms for communicating with the environment external to their human host, and are not easily detectable using the sort of electronic equipment that is typically used to detect and analyze mobile computers), it may be more prudent, effective, and efficient to attempt to entirely mask and conceal the device's existence than to create a honeypot which, in a sense, is purposefully designed to attract attention.[58]

---

[56] *NIST SP 800-53* (2013), p. F–208.
[57] *NIST SP 800-53* (2013), p. F–202.
[58] See Chapter Three of this text for proposed approaches that utilize shielding or jamming in an attempt to conceal the existence of a neuroprosthetic device.

SDLC stage 5: device disconnection, removal, and disposal

The fifth stage in the system development life cycle involves a neuprosthetic device's functional removal from its host-device system and broader supersystem; this may be accomplished through means such as remote disabling of the device or its core functionality, surgical extraction of the device, or the device's physical disassembly or destruction. The stage also includes a device's preparation for reuse or ultimate disposal after removal from its previous human host. The development or execution of security controls in this stage of the SDLC is typically performed by a device's operator or maintenance service provider(s), potentially with the active or passive participation of its human host. In this text, we do not identify any standard detective InfoSec controls as finding their greatest possible relevance during this final stage of the SDLC.

Conclusion

In this chapter, we have reviewed a number of detective security controls for information systems and discussed the implications of applying such controls to neuroprosthetic devices and the larger information systems in which they participate, using the lens of a five-stage system development life cycle as a conceptual framework. In the following chapter, a similar analysis of corrective and compensating controls will be undertaken.

Chapter Eight

Corrective and Compensating Security Controls for Neuroprosthetic Devices and Information Systems

Abstract. This chapter explores the way in which standard corrective and compensating security controls (such as those described in *NIST Special Publication 800-53*) become more important, less relevant, or significantly altered in nature when applied to ensuring the information security of advanced neuroprosthetic devices and host-device systems. Controls are addressed using an SDLC framework whose stages are (1) supersystem planning; (2) device design and manufacture; (3) device deployment; (4) device operation; and (5) device disconnection, removal, and disposal.

Corrective and compensating controls considered include those relating to incident response procedures, mechanisms, and training; error handling capacities; failure mode capacities and procedures; and flaw remediation.

Introduction

In this chapter, we review a range of standard corrective and compensating security controls for information systems and identify unique issues that arise from the perspective of information security, biomedical engineering, organizational management, and ethics when such controls are applied to neuroprosthetic devices and larger information systems that include neuroprosthetic elements. The text applies such security controls without providing a detailed explanation of their workings; it thus assumes that the reader possesses at least a general familiarity with security controls. Readers who are not yet acquainted with such controls may wish to consult a comprehensive catalog such as that found in *NIST Special Publication 800-53, Revision 4*, or *ISO/IEC 27001:2013*.[1]

---

[1] See *NIST Special Publication 800-53, Revision 4: Security and Privacy Controls for Federal Information Systems and Organizations* (2013) and *ISO/IEC 27001:2013, Information technology – Security techniques – Information security management systems – Requirements* (2013).

Approaches to categorizing security controls

Some InfoSec researchers categorize controls as either **administrative** (i.e., comprising organizational policies and procedures), **physical** (e.g., created by physical barriers, security guards, or the physical isolation of a computer from any network connections), or **logical** (i.e., enforced through software or other computerized decision-making).[2] Other sources have historically classified controls as either **management**, **operational**, or **technical** controls. In this volume, we follow the lead of texts such as *NIST SP 800-53*,[3] which has removed from its security control catalog the explicit categorization of such measures as management, operational, or technical controls, due to the fact that many controls incorporate aspects of more than one category, and it would be arbitrary to identify them with just a single category. We instead utilize a classification of such measures as **preventive**, **detective**, or **corrective and compensating** controls. The previous two chapters discussed the first two types of controls, while this chapter investigates the final type.

Role of security controls in the system development life cycle

The corrective and compensating controls discussed in this chapter are organized according to the stage within the process of developing and deploying neuroprosthetic technologies when attention to a particular control becomes most relevant. These phases are reflected in a system development life cycle (SDLC) whose five stages are (1) supersystem planning; (2) device design and manufacture; (3) device deployment in the host-device system and broader supersystem; (4) device operation within the host-device system and supersystem; and (5) device disconnection, removal, and disposal.[4] Many controls relate to more than one stage of the process: for example, the decision to develop a particular control and the formulation of its basic purpose may be developed in one stage, while the details of the control are designed in a later stage and the control's mechanisms are implemented in yet another stage. Here we attempt to locate a control in the SDLC stage in which decisions or actions are undertaken that have the greatest impact on the success

---

[2] Rao & Nayak, *The InfoSec Handbook* (2014), pp. 66-69.

[3] See *NIST SP 800-53* (2013).

[4] A four-stage SDLC for health care information systems is described in Wager et al., *Health Care Information Systems: A Practical Approach for Health Care Management* (2013), a four-stage SDLC for an open eHealth ecosystem in Benedict & Schlieter, "Governance Guidelines for Digital Healthcare Ecosystems" (2015), pp. 236-37, and a generalized five-stage SDLC for information systems in *Governance Guidelines for Digital Healthcare Ecosystems* (2006), pp. 19-25. These are synthesized to create a five-stage SDLC for information systems incorporating brain-computer interfaces in Gladden, "Managing the Ethical Dimensions of Brain-Computer Interfaces in eHealth: An SDLC-based Approach" (2016).

or failure of the given control. This stage-by-stage discussion of corrective and compensating controls begins below.

SDLC stage 1: supersystem planning

The first stage in the system development life cycle involves high-level planning of an implantable neuroprosthetic device's basic capacities and functional role, its relationship to its human host (with whom it creates a biocybernetic host-device system), and its role within the larger 'supersystem' that comprises the organizational setting and broader environment within which the device and its host operate. The development of security controls in this stage of the SDLC typically involves a neuroprosthetically augmented information system's designer, manufacturer, and eventual institutional operator. Such controls are considered below.

A. Developing incident response procedures

1. Incident response teams

The use of dedicated organizational incident response teams[5] or services that proactively respond to an ongoing incident (e.g., by physically locating the host of a neuroprosthetic device, assessing his or her condition, and providing containment and recovery services) may be especially necessary in the case of a neuroprosthesis whose anomalous functioning may render its host incapacitated and unable to respond to an incident himself or herself.

2. Incident reporting methods

Various incident reporting[6] complications can arise with neuroprosthetically augmented information systems. For example, in the case of a human host who does not even realize that he or she has been implanted with a neuroprosthetic device, the host might discern that he or she is undergoing some unusual experience but would not associate it with the device and may have no ability to report the incident to the device's operator.[7]

3. Design of fail-safe procedures for the supersystem

The design and implementation of effective fail-safe procedures[8] is essential for ensuring information security for advanced neuroprosthetic devices and host-device systems, especially those with critical health impacts for a

---

[5] *NIST SP 800-53* (2013), p. F–108.

[6] *NIST SP 800-53* (2013), p. F–107.

[7] For the possibility that human hosts might unwittingly be implanted, e.g., with certain kinds of RFID devices, see Gasson, "Human ICT Implants: From Restorative Application to Human Enhancement" (2012).

[8] *NIST SP 800-53* (2013), p. F–233.

device's host. Such procedures may require, for example, that the host or operator of a neuroprosthetic device receive a clear automated alert upon the failure or impending failure of critical device components, systems, or processes, along with explicit instructions of steps that should be taken. In the event of some failures by certain kinds of neuroprosthetic devices, a device's host may have only minutes or seconds in which to execute specified fail-safe procedures before the failure incapacitates the host or otherwise renders him or her unable to take additional action. Other specific fail-safe procedures may include enabling mechanisms that will allow emergency medical personnel to access a neuroprosthetic device, initiating the backup of key information maintained in volatile memory, or releasing particular biochemical agents to stimulate a specific response in the body of the device's host.[9]

B. Planning of incident response training

1. Designing incident response training

Incident response training[10] is especially important for the human host of an advanced neuroprosthesis, insofar as an incident relating to such a device does not compromise or damage some external system like a desktop computer or smartphone but may actually compromise the host's own biological and cognitive processes. Once an incident is underway, a device's host may only have a very limited time in which to react and carry out response measures before losing consciousness or losing control over his or her own volition, memory, or other mental processes. Specialized incident response training can help ensure that a device's host recognizes and responds to an ongoing incident in a timely and effective manner.

2. Planning of automated training environments

An organization or individual may utilize "automated mechanisms to provide a more thorough and realistic incident response training environment."[11] With some kinds of neuroprosthetic devices, automated training environments that are governed by artificially intelligent systems may be needed to accurately simulate or replicate the activity of adversaries whose capacities and techniques exceed the limits of what can be possessed or performed by an unaugmented human adversary.

---

[9] See Chapter Three of this text for a discussion of failure modes for neuroprosthetic devices and the need to provide adequate access to emergency medical personnel when a device's host is experiencing a medical emergency.

[10] *NIST SP 800-53* (2013), p. F–103.

[11] *NIST SP 800-53* (2013), p. F–104.

SDLC stage 2: device design and manufacture

The second stage in the system development life cycle includes the design and manufacture of a neuroprosthetic device and other hardware and software that form part of any larger information system to which the device belongs. The development of security controls in this stage of the SDLC is typically performed by a device's designer and manufacturer, potentially with instructions or other input from the system's eventual operator. Such controls are considered below.

A. Error handling capacities

1. Design of error handling procedures

Error messages generated by information systems should generally provide the kinds of information needed for organizational personnel to identify and remedy the source of the error without providing information that could be used by adversaries to either directly compromise a system or indirectly learn more about its functioning.[12] In the case of neuroprosthetic devices with critical health impacts, error messages may need to be presented not only through internal sensory or cognitive processes to a device's host or through organizational information systems to the device's operator but potentially also to emergency medical personnel who previously had no connection to the host or the host's organization but who happened to be in the vicinity of a host, are diagnosing and treating him or her for some health emergency, and may not (yet) have full access to the neuroprosthetic device's components or processes.[13]

2. Designing automated responses to integrity violations

Care must be taken that any automated responses to integrity violations[14] detected within a neuroprosthetic device do not cause physical or psychological harm to the device's host or others. In some circumstances, automated responses may need to be delayed in order to prevent a device from malfunctioning or failing when it is being used by its host or operator to perform an

---

[12] *NIST SP 800-53* (2013), p. F–230.

[13] Chapter Three of this text considers many proposed technological approaches for granting emergency medical personnel access to a neuroprosthetic device in cases when its human host is experiencing a medical emergency. An underappreciated aspect of such situations is the fact that even if emergency medical personnel have the technological means by which to gain access to a particular neuroprosthetic device, this in no way guarantees that they will have the expertise in computer science, information technology, biomedical engineering, or cybernetics that may be required in order to quickly diagnose the device's status and functioning, alter its configuration, and perhaps even reprogram it in order to yield specific positive outcomes for and impacts on the host's biological organism.

[14] *NIST SP 800-53* (2013), p. F–226.

urgent task (e.g., with potentially critical health impacts for the device's host). In other circumstances, an automated response may *need* to take place instantaneously in order to protect the device's host or operator from some critical health impact.

3. Integration of incident detection and response

In the case of certain neuroprostheses – for example, some utilizing a physical neural network that is broadly interconnected with the neural circuitry of the brain of the devices' host and whose operating system or applications are partly or wholly stored within the biological structures and cognitive processes of the host's brain – the detection of integrity violations and the response to them may inherently be closely integrated,[15] insofar as the same artificial neurons that receive input through their synthetic dendrites that allows a violation to be detected will also be involved in transmitting output through their synthetic axons to the connected natural biological neurons in an effort to remedy the integrity violation.

B. Design of failure mode capacities and procedures

1. Design of the capacity to fail in a known state

Neuroprosthetic devices may be designed so that they fail (at least in the case of some kinds of failures) in a known state that preserves information about the devices' final pre-failure state.[16] It can be especially helpful for a neuroprosthesis to be designed to fail in a known secure state in cases when an implanted neuroprosthetic device cannot easily be inspected or otherwise immediately accessed to externally determine or confirm its failure state.[17]

2. Design of the capacity to fail secure

It is important that advanced neuroprostheses be designed to fail securely in the case of a failure of one of a device's boundary protection systems or components;[18] however, the basic concept of 'secure failure' may take on an unusual form in this context. Under normal circumstances, secure failure implies that after the failure of a boundary protection, information will be unable to either enter or leave a system until the failure has been remedied. In the case of neuroprosthetic devices, the state of secure failure may require

---

[15] Regarding integrated incident detection and response, see *NIST SP 800-53* (2013), p. F–226.

[16] *NIST SP 800-53* (2013), p. F–202.

[17] See Chapter Three of this text for a discussion of issues relating to the lack of physical access to neuroprosthetic devices after their implantation.

[18] *NIST SP 800-53* (2013), pp. F–191-92.

that some information be able to enter and leave a device (or host-device system) in order to avoid causing direct physical or psychological harm to a device's human host.

Hansen and Hansen argue that in general, implantable medical devices should be designed in such a way that if an entire device, its individual components, or the larger system's security controls fail, they will 'fail open' in a way that allows rather than prevents the flow of information and access to the device, "since it is almost always better to give possibly-inappropriate access if the alternative is death or disability [...]."[19] In the case of a particular advanced neuroprosthetic device, it must be carefully investigated and determined whether failing into a state that is 'open' or 'closed' is more likely to lead to severe harm (or even death) for the device's host or operator under different kinds of possible circumstances. The types of information that should be allowed to enter or leave a device during failures should be determined by a device's designer in collaboration with physicians, psychologists, and biomedical engineers who possess relevant expertise about the potential physical and psychological impacts of device failure and a loss of information flow on a device's human host.

3. Design and installation of standby or backup components

Often the failure of a component triggers the automatic or manual transfer of the component's responsibilities to a standby component that was already in place and ready to be activated.[20] In the case of implanted neuroprosthetic devices whose ability to communicate with external systems cannot be reliably guaranteed, whose functioning depends on direct physical access to biological structures or processes within their host's body, or which cannot be easily manually accessed for repair or replacement, it may not be possible to utilize standby components that are located externally to a host's body: any standby components may need to be implanted into a host's body at the same time as the primary neuroprosthetic device or may need to be directly incorporated into the structure of that primary neuroprosthesis itself.

4. Design of the failover to standby or backup systems

The automatic switchover to an alternate system after the failure of an entire information system typically requires that mirrored systems or alternate processing sites[21] have already been established and adequately prepared and maintained in advance of the moment when the failure occurs. If by 'system' we understand an entire neuroprosthetic device, the unexpected failure of such a system may cause significant physical or psychological harm to the

---

[19] Hansen & Hansen, "A Taxonomy of Vulnerabilities in Implantable Medical Devices" (2010).
[20] *NIST SP 800-53* (2013), p. F–231.
[21] *NIST SP 800-53* (2013), p. F–231.

device's host or operator, especially if the device has critical health impacts. In some cases, it may not be possible to install an alternate system at the same time as implantation of the primary neuroprosthesis due to practical constraints such as space or power limitations or the fact that key biological structures and processes within the body of the primary device's host can interface with at most one device of that kind at a time.

In the case of 'failure' of an entire host-device system (e.g., through the incapacitation or death of a device's human host), there may not be any possibility of failover to an alternate information system, insofar as it is not feasible to 'mirror' in a synthetic external information system such traits as the unique physical, legal, and ontological identity or continuity of consciousness and agency of a particular human being, no matter how closely the external system may mimic some other traits displayed by the person (such as his or her genotype, physical appearance, or even the contents of his or her memory). While the concept of 'uploading' key information relating to a human being into an information system or creating physical or virtual copies of critical physical components or processes of the person has been proposed by some transhumanists and much debated[22] – and such techniques could indeed be said to provide a limited 'failover capability,' if the only goal is to preserve (partial and potentially inaccurate) records of some aspects of a human being's physical structure at a given point in time or of the person's past behavior – such mechanisms do not effect the continuation of a human being's essence or existence in any robust sense.

Perhaps the only way in which a neuroprosthetic device or neurocybernetic system could allow the continuation of the existence of a 'human being' through failover to an alternate information system would be if the individual were not a natural biological human being to begin with (in the way that the expression is traditionally understood) but were rather a simulacrum that was already, in some sense, a copy or representation without an original.[23] If the traditional understanding of the concept of a 'human being' were someday to be expanded or transformed to such an extent that an information system, virtual entity, software program, or instantiation of patterns within a neural network could be considered a 'human being' simply because it displays certain human-like characteristics or contains information derived from human

---

[22] Regarding such matters, see Koene, "Embracing Competitive Balance: The Case for Substrate-Independent Minds and Whole Brain Emulation" (2012); Proudfoot, "Software Immortals: Science or Faith?" (2012); Pearce, "The Biointelligence Explosion" (2012); Hanson, "If uploads come first: The crack of a future dawn" (1994); and Moravec, *Mind Children: The Future of Robot and Human Intelligence* (1990).

[23] See, e.g., Baudrillard, *Simulacra and Simulation* (1994), for a discussion of such issues from a philosophical perspective.

beings – without requiring that such information be housed within or accessed through a particular unique biological substrate – then it would be possible to imagine the preservation and continuation of an entire host-device system through failover to an entirely disjoint alternate system. However, such 'preservation' or 'continuation' is not at all the sort of preservation and continuation of personal consciousness, agency, and physical and noetic identity that a human being possessing a neuroprosthetic device would generally seek and which it may be the legal and ethical responsibility of the device's operator to ensure.[24]

5. Design of automatic device shutdown on audit failure

For some kinds of advanced neuroprostheses it may be essential that a device automatically shut down if its audit-processing ability is compromised (e.g., due to a hardware error or reaching the audit storage capacity), in order to avoid the possibility that the loss of audit-processing ability might allow the device to inflict physical or psychological harm on its host or others.[25] In other cases, a device may be required to continue operating after its audit-processing ability has been compromised and until it can be restored, due to the fact that the abrupt cessation of operations could inflict harm on its host or others.

C. Design of incident response mechanisms

1. Design of automated incident handling procedures

Relying on automated incident handling processes[26] to perform functions of incident detection, containment, and eradication may be hazardous if the execution of such functions is able or likely to directly or indirectly cause physical or psychological harm to a neuroprosthetic device's human host; an automated incident response system may not always recognize the effect that its efforts are inadvertently having on the device's host. On the other hand, in other situations, relying on an automated incident response system may be *less* likely to result in harm to a device's host than having human agents directly control the response, if the system can be trained to respond with a greater degree of speed, accuracy, and effectiveness than a human agent.

---

[24] See Proudfoot (2012). See also Chapter Three of this text for a discussion of the need to protect the personal identity, autonomy, agency, and sapient self-awareness of a neuroprosthetic device's human host.

[25] *NIST SP 800-53* (2013), p. F–44.

[26] *NIST SP 800-53* (2013), p. F–105.

2. Design of dynamic reconfiguration as an incident response

Neuroprosthetic systems may be designed to dynamically reconfigure themselves either as a routine matter (in order to prevent potential attacks) or in response to an ongoing incident, in order "to stop attacks, to misdirect attackers, and to isolate components of systems, thus limiting the extent of the damage from breaches or compromises."[27] In the case of neuroprosthetic devices utilizing biological components or physical neural networks, some degree of 'dynamic reconfiguration' may be continuously taking place.[28]

3. Design of dynamic information flow control as incident response

Particular kinds of information flow controls might be automatically enabled or disabled[29] if, for example, a neuroprosthetic device detects that its human host is experiencing a medical emergency or if the device's operator determines that the host is entering a situation in which specialized information flows are warranted.

4. Design of backup controls as incident response

A neuroprosthetic device may possess backup security controls (e.g., alternate methods for user authentication) that become active only if the device's primary controls have been compromised.[30]

5. Designing automated responses to denial of service attacks

Denial of service attacks[31] may take on new forms in the cases of some advanced neuroprosthetic devices. For example, a sensory neuroprosthesis such as an artificial eye could potentially be subjected to a successful denial of service attack by exposing it to an intense light source or array of light

---

[27] *NIST SP 800-53* (2013), p. F–105.

[28] For example, for a discussion of the ways in which long-term memories stored within the human brain can undergo changes in their nature and storage over time, see Dudai, "The Neurobiology of Consolidations, Or, How Stable Is the Engram?" (2004). In a sense, memories stored within the brain's natural biological neural networks that undergo such changes (even if subtle ones) over time might be thought of as loosely analogous to metamorphic or polymorphic malware: a stored memory's ongoing dynamic reconfiguration may make it more difficult for adversaries to target that particular memory for alteration, manipulation, or deletion, if the storage location and identifying characteristics of the memory are not entirely stable. For factors that may either enhance or limit the dynamic reconfiguration of memories stored within the human brain, see, e.g., the discussion of holographic brain models in Longuet-Higgins, "Holographic Model of Temporal Recall" (1968); Westlake, "The possibilities of neural holographic processes within the brain" (1970); Pribram, "Prolegomenon for a Holonomic Brain Theory" (1990); and Pribram & Meade, "Conscious Awareness: Processing in the Synaptodendritic Web – The Correlation of Neuron Density with Brain Size" (1999).

[29] See *NIST SP 800-53* (2013), p. F–15.

[30] Regarding such controls, see *NIST SP 800-53* (2013), p. F–89.

[31] *NIST SP 800-53* (2013), p. F–187.

sources that overwhelms, confuses, or blocks its ability to gather desired information from the environment. Denial of service attacks can also take the form of **resource depletion attacks** that attempt to exhaust the internal battery or other power source of an implantable neuroprosthesis by subjecting it to an unending string of wireless access requests from some external system: even if the neuroprosthetic device successfully rejects all of the unauthorized access requests, the work of responding to and verifying each request can quickly exhaust the device's battery and disable it.[32]

6. Determining the response to unsuccessful logon attempts

A control that automatically locks an account or delays the next logon prompt after a specified number of consecutive unsuccessful logon attempts[33] may be hazardous in emergency situations in which access to a device is needed immediately in order to prevent physical or psychological harm to its human host or others (and which may be precisely the sort of situation in which ongoing stress or physical impairment may cause the host's logon attempts to be unsuccessful).[34]

7. Design of the automatic wiping of a device in response to unsuccessful logon attempts

A control that automatically purges data from a neuroprosthetic device after a certain number or type of unsuccessful logon attempts may be desirable in order to preserve the confidentiality and possession of sensitive data stored within the device.[35] On the other hand, such countermeasures may be legally or ethically impermissible in situations in which they would cause physical or psychological damage to a device's host or to others.

8. Coordination of incident response processes with component suppliers

In some cases, successfully responding to an incident impacting an advanced neuroprosthesis may require coordination between the device's designer, manufacturer, OS and application developers, provider, installer, operator, and human host.[36]

---

[32] See the discussion of threats in Chapter Two of this text for more about resource depletion attacks.

[33] *NIST SP 800-53* (2013), p. F–21.

[34] See Chapter Three of this book for alternative methods for preventing resource depletion attacks involving a string of unsuccessful logon attempts, and see Chapter Two for a basic description of resource depletion attacks.

[35] Regarding the automatic wiping of devices, see *NIST SP 800-53* (2013), p. F–21.

[36] *NIST SP 800-53* (2013), p. F–106.

SDLC stage 3: device deployment in the host-device system and broader supersystem

The third stage in the system development life cycle includes the activities surrounding deployment of a neuroprosthetic device in its human host (with whom it forms a biocybernetic host-device system) and the surrounding organizational environment or supersystem. The development or implementation of security controls in this stage of the SDLC is typically performed by a device's operator with the active or passive participation of its human host. Such controls are considered below.

A. Activation of incident response mechanisms

1. Automated intrusion detection and response

The use of automated mechanisms to detect intrusions into a device or the surrounding body of its human host and to initiate particular response actions[37] must be undertaken carefully, insofar as some forms of intrusion (e.g., medical procedures) may be done with the host's consent and at his or her direct request, and automated responses could potentially cause physical or psychological harm if initiated while the host were in the midst of performing or undergoing some critical activity or otherwise at a time not desired by the host.

2. Detection and blocking of threatening outgoing communications

If viewed solely from the perspective of a neuroprosthetic device, **extrusion detection**[38] is essential for preventing potentially harmful traffic from passing from the device into the cognitive processes or biological systems of the device's human host. However, if viewed from the perspective of the larger host-device system, such controls are not extrusion detection but internal controls; true extrusion detection would attempt to detect and prevent, for example, the use of a neuroprosthesis by its host or operator to conduct unauthorized denial of service attacks, the dissemination of malware, illegal surveillance, or other illicit actions targeted at external systems or individuals.

B. Testing of incident response procedures

1. Use of simulated events for incident response testing

Incidents may be especially easy to simulate[39] for the host of a neuroprosthetic device when the device already creates a virtual or augmented reality

---

[37] Regarding automated intrusion detection and response mechanisms, see *NIST SP 800-53* (2013), p. F–131.

[38] *NIST SP 800-53* (2013), p. F–190.

[39] *NIST SP 800-53* (2013), p. F–104.

for the host by supplying artificial sense data. At the same time, though, it may potentially be difficult for the hosts of such devices to distinguish simulated events from actual ones.[40]

2. Automated testing of incident response processes

Automated testing[41] may be necessary for neuroprosthetic devices that cannot directly be accessed by technologies or activities controlled by human agents but which may, for example, be accessible and potentially vulnerable to attacks utilizing nanorobotic swarms or other automated systems.

SDLC stage 4: device operation within the host-device system and supersystem

The fourth stage in the system development life cycle includes the activities occurring after a neuroprosthetic device has been deployed in its production environment (comprising its host-device system and broader supersystem) and is undergoing continuous use in real-world operating conditions. The development or execution of security controls in this stage of the SDLC is typically performed by a device's operator and maintenance service provider(s) with the active or passive participation of its human host. Such controls are considered below.

A. Flaw remediation

1. Centralized management of flaw remediation

Some kinds of 'flaws' detected in the functioning or operation of a neuroprosthetic device may be flaws not in the physical device or its software but in the structure and behavior of the larger host-device system in which it participates; in such circumstances, detection and remediation[42] of the flaw may depend largely on the individual capacities and action of the device's human host rather than the organizations responsible for designing, manufacturing, providing, or operating such neuroprostheses.[43]

2. Establishing deadlines and benchmarks for flaw remediation

In the case of some kinds of neuroprosthetic devices that interact with or support biological processes critical to the health of their human host, both

---

[40] See Chapter Four of this text for a discussion of the distinguishability of neuroprosthetically supplied information as an information security goal and attribute for neuroprosthetic devices.
[41] *NIST SP 800-53* (2013), p. F–104.
[42] Regarding centralized management of flaw remediation, see *NIST SP 800-53* (2013), p. F–216.
[43] See Chapter Three of this text for a discussion of the distinction between a neuroprosthetic device and its host-device system.

legal, ethical, and operational considerations may dictate that a flaw must be corrected immediately upon its detection;[44] the fact that an organization works "as quickly as possible" to resolve the problem may not absolve it of responsibility for damage that occurs as a result. This is especially true in the case of devices that cannot easily be recalled, removed, or replaced if a flaw is discovered after a device has been implanted in its human host.[45]

3. Automatic updating of firmware and software to eliminate vulnerabilities

Allowing the automatic downloading, installation, and execution of operating system or application updates[46] by neuroprosthetic devices should only be undertaken after careful consideration, especially for devices with critical health impacts for their human host. The fact that operating system or application updates have undergone beta testing in a simulated development environment or with a limited number of host-device systems prior to their widespread public release may not ensure that the updates will not cause severe and unexpected negative impacts on the functioning of some implanted neuroprosthetic devices and harm for their human hosts, given the fact that the functioning of individual neuroprostheses may vary greatly depending on the unique nature of each device's physical interface with the neural circuitry of its human host and the nature of the host's cognitive patterns and activity.

B. Incident response

1. Use of detonation chambers for execution of suspicious code

The ability to implement a detonation chamber (or 'dynamic execution environment') within a neuroprosthetic device may be limited by the fact that some malicious code or applications may not be inherently (or obviously) harmful in themselves but only when allowed to interact with or be run by the cognitive processes of a particular human host. If it is not possible to simulate a host's cognitive processes with sufficient richness and accuracy in some artificial dynamic execution environment, then carrying out actions such as executing suspicious programs, opening suspicious email attachments, or visiting suspicious websites[47] within the detonation chamber may

---

[44] Regarding deadlines and benchmarks for flaw remediation, see *NIST SP 800-53* (2013), p. F–216.

[45] See Chapter Three of this text for a discussion of how the concept of zero-day vulnerabilities and attacks relates to neuroprosthetic devices – and especially to those possessing critical health impacts for their human host.

[46] *NIST SP 800-53* (2013), p. F–216.

[47] *NIST SP 800-53* (2013), p. F–214.

not reveal the harmful effects that the same actions would have when performed by a neuroprosthetic device within a particular host-device system.[48]

2. Information spillage response

Care must be exercised in defining information spillage[49] with regard to neuroprosthetically augmented information systems and formulating spillage responses. For example, imagine that a human host possesses a neuroprosthetic device that stores information on flash memory that the host can access and 'play back' to his or her conscious awareness through sensory systems.[50] The host may have access and authorization to view classified information stored on the device, but viewing the information would create a (perhaps not entirely accurate) additional copy of the information in the natural biological long-term memory system within the host's brain; this could potentially be considered an information spillage. A similar situation would occur if a person were authorized to read printouts of classified information while in a secured location but not to transfer the information to a digital storage system; if the person's long-term memory processes were augmented with engram-storing mnemoprostheses, simply reading the documents in an authorized manner could result in the production of an unauthorized copy of the information within the neuroprosthetic system. In such situations, manual or automated responses that seek to take 'corrective action' to contain and eradicate spillage within the systems that have been 'contaminated'[51] by the information spillage have the potential to cause physical and psychological damage to a device's human host.

3. System recovery and reconstitution

The exact process of system recovery and reconstitution for a neuroprosthetically augmented information system will depend on the nature of the failure that has made the recovery process necessary and the extent of damage that the system and its stored information may have suffered. In the case

---

[48] Practical difficulties with implementing a detonation chamber within a neuroprosthetic device itself could arise either from the nature of the device's computing platform (e.g., it may be difficult or impossible to implement such an environment within a neuroprosthesis that does not execute traditional programs but instead processes information using a physical – and perhaps biological – neural network) or simply from limitations on the processing power, storage capacity, or power supply of a device's internal computer. See the device ontology in Chapter One of Gladden, *Neuroprosthetic Supersystems Architecture* (2017), for a discussion of such considerations.

[49] *NIST SP 800-53* (2013), p. F–109.

[50] For the idea of such sensory playback capabilities, see Merkel et al., "Central Neural Prostheses" (2007); Robinett, "The consequences of fully understanding the brain" (2002); and McGee, "Bioelectronics and Implanted Devices" (2008), p. 217.

[51] *NIST SP 800-53* (2013), p. F–109-110.

of some kinds of neuroprostheses (e.g., those utilizing a complex physical neural network), it may theoretically be possible to scan and record the state of the entire device at a single moment in time – thus creating a backup file – however, there may be no mechanism available for restoring the system to a previous state by overwriting the device's current state with the information contained in the backup file.[52]

SDLC stage 5: device disconnection, removal, and disposal

The fifth stage in the system development life cycle involves a neuroprosthetic device's functional removal from its host-device system and broader supersystem; this may be accomplished through means such as remote disabling of the device or its core functionality, surgical extraction of the device, or the device's physical disassembly or destruction. The stage also includes a device's preparation for reuse or ultimate disposal after removal from its previous human host. The development or execution of security controls in this stage of the SDLC is typically performed by a device's operator or maintenance service provider(s), potentially with the active or passive participation of its human host. In this text, we do not identify any standard detective InfoSec controls as finding their greatest possible relevance during this final stage of the SDLC.

Conclusion

In this chapter, we have reviewed a number of standard corrective and compensating security controls for information systems and discussed the implications of applying such controls to neuroprosthetic devices and the larger information systems in which they participate, using the lens of a five-stage system development life cycle as a conceptual framework. This concludes our investigation of preventive, detective, and corrective or compensating controls and their relationship to neuroprosthetic devices and neuroprosthetically augmented information systems.

---

[52] Regarding information system recovery and reconstitution, see *NIST SP 800-53* (2013), pp. F–87-88.

Appendix

Appendix

Information Security Concerns as a Catalyst for the Development of Implantable Cognitive Neuroprostheses[1]

Abstract. Standards like the *ISO 27000* series, *IEC/TR 80001*, *NIST SP 1800*, and FDA guidance on medical device cybersecurity define the responsibilities that manufacturers and operators bear for ensuring the information security of implantable medical devices. In the case of implantable cognitive neuroprostheses (ICNs) that are integrated with the neural circuitry of their human hosts, there is a widespread presumption that InfoSec concerns serve only as limiting factors that can complicate, impede, or preclude the development and deployment of such devices. However, we argue that when appropriately conceptualized, InfoSec concerns may also serve as drivers that can spur the creation and adoption of such technologies. A framework is formulated that describes seven types of actors whose participation is required in order for ICNs to be adopted; namely, their 1) producers, 2) regulators, 3) funders, 4) installers, 5) human hosts, 6) operators, and 7) maintainers. By mapping onto this framework InfoSec issues raised in industry standards and other literature, it is shown that for each actor in the process, concerns about information security can either disincentivize or incentivize the actor to advance the development and deployment of ICNs for purposes of therapy or human enhancement. For example, it is shown that ICNs can strengthen the integrity, availability, and utility of information stored in the memories of persons suffering from certain neurological conditions and may enhance information security for society as a whole by providing new tools for military, law enforcement, medical, or corporate personnel who provide critical InfoSec services.

Introduction

Developments in the field of neuroprosthetics are occurring at a rapid pace. Among the most revolutionary technologies are implantable cognitive neuroprostheses (ICNs) that are housed permanently within a human host's

[1] This text was originally published as Gladden, Matthew E., "Information Security Concerns as a Catalyst for the Development of Implantable Cognitive Neuroprostheses," in *9th Annual EuroMed Academy of Business (EMAB) Conference: Innovation, Entrepreneurship and Digital Ecosystems (EUROMED 2016) Book of Proceedings*, edited by Demetris Vrontis, Yaakov Weber, and Evangelos Tsoukatos, pp. 891-904; Engomi: EuroMed Press, 2016.

body and which interact with the brain to regulate or enhance cognitive processes relating to memory, emotion, imagination, belief, and conscious awareness.

If such devices fail to function as intended, they can have a severe negative impact on the psychological and physical well-being of their human hosts. While information security (InfoSec) experts have begun formulating approaches to safeguarding these devices against computer viruses, cyberattacks, communication glitches, power outages, user authentication errors, and other problems that could disrupt their functioning, it is commonly presumed that InfoSec concerns represent a significant obstacle to the broader adoption of such technologies. Almost no consideration has been given to the possibility that InfoSec concerns might also create compelling reasons *in favor of* developing and deploying ICNs within society.

In this text, a conceptual framework is formulated which demonstrates that at each step in the process of creating and adopting ICNs, it is possible for InfoSec-related concerns to either impede the process or drive it forward. Before considering that framework, we can review the state of ICNs and industry standards for information security, especially as it applies to implantable medical devices.

Background and foundations

Overview of implantable cognitive neuroprosthetics

A neuroprosthesis can be understood as "a technological device that is integrated into the neural circuitry of a human being."[2] Such neuroprostheses can be sensory, motor, or cognitive in nature.[3] In this text we focus on cognitive neuroprostheses – experimental devices that enhance, regulate, replace, or otherwise participate in cognitive processes and phenomena[4] such as memory,[5] emotion,[6] personal identity and agency,[7] and consciousness.[8]

---

[2] Gladden, *The Handbook of Information Security for Advanced Neuroprosthetics* (2015), p. 21; Lebedev, "Brain-Machine Interfaces: An Overview" (2014).

[3] Lebedev (2014).

[4] Gladden (2015), pp. 26-27.

[5] Han et al., "Selective Erasure of a Fear Memory" (2009); Ramirez, "Creating a False Memory in the Hippocampus" (2013).

[6] Soussou & Berger, "Cognitive and Emotional Neuroprostheses" (2008).

[7] Van den Berg, "Pieces of Me: On Identity and Information and Communications Technology Implants" (2012).

[8] Kourany, "Human Enhancement: Making the Debate More Productive" (2013); Claussen & Hofmann, "Sleep, Neuroengineering and Dynamics" (2012).

Such devices are still in their early experimental stages; however, it is anticipated that they will eventually be used to treat a range of conditions such as anxiety disorders, emotional disorders, addictions, Alzheimer's disease, and other memory disorders[9] as well as to enhance cognitive capacities like memory and alertness beyond their natural limits.[10]

Some neuroprosthetic technologies comprise large and sessile pieces of non-invasive equipment (e.g., fMRI machines) that are permanently housed in dedicated medical facilities and can only be used at those locations. Other neuroprosthetic technologies involve prostheses that are physically integrated into the biological organism of a human host but have an interface with the external environment; still others are implants which, after their surgical insertion, are entirely concealed within the body of a human host (often within the brain) and may remain there throughout the rest of their host's lifetime.[11] In this text we focus on implantable cognitive neuroprosthetic (ICNs), which display unique InfoSec characteristics because they: 1) are often deeply integrated with the biological neural network of their human host's brain, creating the possibility of severe psychological or physical harm (including death) if they are compromised or fail to function as intended; 2) must rely on wireless communication to interact with external health information systems and receive instructions and software updates; and 3) are highly mobile devices that enter a diverse range of unpredictable and unsecure environments as their host goes about his or her daily life.[12]

---

[9] See Ansari et al., "Vagus Nerve Stimulation: Indications and Limitations" (2007); Merkel et al., "Central Neural Prostheses" (2007); Stieglitz, "Restoration of Neurological Functions by Neuroprosthetic Technologies: Future Prospects and Trends towards Micro-, Nano-, and Biohybrid Systems" (2007); Soussou & Berger (2008); Van den Berg (2012); and Gladden (2015), pp. 22-26.

[10] See Spohrer, "NBICS (Nano-Bio-Info-Cogno-Socio) Convergence to Improve Human Performance: Opportunities and Challenges" (2002); McGee, "Bioelectronics and Implanted Devices" (2008); Brunner & Schalk, "Brain-Computer Interaction" (2009); Koops & Leenes, "Cheating with Implants: Implications of the Hidden Information Advantage of Bionic Ears and Eyes" (2012); Kourany (2013); Rao et al., "A direct brain-to-brain interface in humans" (2014); Warwick, "The Cyborg Revolution" (2014); and Gladden (2015), pp. 26-28.

[11] Gladden (2015), pp. 28-29.

[12] *ISO 27799:2008, Health informatics – Information security management in health using ISO/IEC 27002:2013, Information technology – Security techniques – Code of practice for information security controls* (2013), p. 47; *NIST Special Publication 1800-1: Securing Electronic Health Records on Mobile Devices (Draft)* (2016), Part a, p. 1; *Content of Premarket Submissions for Management of Cybersecurity in Medical Devices: Guidance for Industry and Food and Drug Administration Staff* (2014), p. 4; Gladden (2015), pp. 62-65.

Fundamental principles of information security (InfoSec)

Information security is an interdisciplinary field whose goal has traditionally been to ensure the *confidentiality, integrity,* and *availability* of information.[13] This notion of a 'CIA Triad' has been expanded through Parker's vision of safeguarding the three additional attributes of the *possession, authenticity,* and *utility* of information.[14] However, a neuroprosthetic device is not a conventional computerized information system; as an instrument integrated into the neural circuitry of its human host, it becomes part of the personal 'information system' that comprises the host's mind and body and which possesses a unique legal and moral status. As a result, ensuring information security for a neuroprosthesis also entails safeguarding the three additional attributes of *distinguishability,* or the possibility of differentiating information according to its nature or origin (e.g., the ability to recognize which of the thoughts experienced in one's mind are 'one's own' and which, if any, are being generated or altered by a neural implant); *rejectability,* or the ability of a host-device system to purposefully exclude particular information from the host's conscious awareness (i.e., the freedom *not* to recall certain memories or entertain particular thoughts at a given moment); and *autonomy,* or the ability of a host-device system to exercise its own agency in the processing of information (i.e., the ability to arrive at a decision through the use of one's own cognitive processes and of one's own volition, without the contents of that decision being manipulated or determined by some external agent).[15] In the case of a neuroprosthetic device integrated with the neural circuitry of its human host, information security thus involves not only securing all electronic data stored in or processed by the device but also ensuring the integrity of the thoughts, memories, volitions, emotions, and other informational processes and content of the natural biological portions of the host's mind in the face of a full range of vulnerabilities and threats including electronically, biologically, and psychologically based attacks.[16]

A key mechanism for promoting information security is the implementation of administrative, physical, and logical security controls.[17] This does not simply involve the installation of antivirus software but rather the creation and effective implementation of a comprehensive program of risk management.[18]

---

[13] Rao and Nayak, *The InfoSec Handbook* (2014), pp. 49-53; *NIST SP 1800-1* (2016), Part b, p. 9; "Security Risk Assessment Framework for Medical Devices" (2014).

[14] Parker, "Toward a New Framework for Information Security" (2002); Parker, "Our Excessively Simplistic Information Security Model and How to Fix It" (2010).

[15] Gladden (2015), pp. 138-42.

[16] Gladden (2015), pp. 40-57; Denning et al., "Neurosecurity: Security and Privacy for Neural Devices" (2009).

[17] Rao and Nayak (2014), pp. 66-69.

[18] *NIST SP 800-33* (2001), p. 19.

InfoSec standards of relevance to implantable cognitive neuroprosthetics

Widely utilized standards that help organizations design and implement best practices for information security include the *ISO 27000* series that defines requirements for InfoSec management systems or ISMSes[19] and a code of practice for InfoSec controls.[20] Similarly, NIST standards address risk management and InfoSec life cycles,[21] InfoSec practices for managers,[22] and security and privacy controls.[23]

Beyond these generic InfoSec standards, national and international bodies are increasingly developing specialized standards relating to health care data and medical devices. For example, ISO has published standards and other resources relating to InfoSec for remotely maintained medical devices and information systems,[24] IT networks that incorporate medical devices,[25] and InfoSec management in the field of health care.[26] In 2015, the NIST issued a draft publication on information security for health records stored or processed on mobile devices.[27] The US Food and Drug Administration has issued guidance relating to cybersecurity for medical devices utilizing off-the-shelf software[28] and to the premarket[29] and postmarket[30] management of cybersecurity for medical devices. Industry organizations such as the Medical Device Privacy Consortium have proposed their own InfoSec standards.[31]

---

[19] *ISO/IEC 27001:2013, Information technology – Security techniques – Information security management systems – Requirements* (2013).

[20] *ISO/IEC 27002* (2013).

[21] *NIST Special Publication 800-37, Revision 1: Guide for Applying the Risk Management Framework to Federal Information Systems: A Security Life Cycle Approach* (2010).

[22] *NIST Special Publication 800-100: Information Security Handbook: A Guide for Managers* (2006).

[23] *NIST Special Publication 800-53, Revision 4: Security and Privacy Controls for Federal Information Systems and Organizations* (2013).

[24] *ISO/TR 11633-1:2009, Health informatics – Information security management for remote maintenance of medical devices and medical information systems – Part 1: Requirements and risk analysis* (2009).

[25] See the *IEC 80001: Application of risk management for IT-networks incorporating medical devices* series (2010-15).

[26] *ISO 27799* (2008).

[27] *NIST SP 1800-1* (2015).

[28] *Guidance for Industry - Cybersecurity for Networked Medical Devices Containing Off-the-Shelf (OTS) Software* (2005).

[29] *Content of Premarket Submissions* (2014).

[30] *Postmarket Management of Cybersecurity in Medical Devices: Draft Guidance for Industry and Food and Drug Administration Staff* (2016).

[31] "Security Risk Assessment Framework for Medical Devices" (2014).

INFORMATION SECURITY CONCERNS THAT MAY...

		Disincentivize Participation in ICNs' Development and Adoption	Incentivize Participation in ICNs' Development and Adoption
Enable Creation	Producers	• Liability for design defects that undermine InfoSec (e.g., that allow psychological damage to end users)	• ICNs might profitably be produced to enhance InfoSec of specialized users (e.g., military) and medical patients
	Regulators	• InfoSec vulnerabilities and possibilities of catastrophic damage may outweigh any benefits of use	• Well-regulated use of ICNs by agencies and health services may enhance InfoSec of society as a whole
Enable Implantation	Funders	• Cost of ensuring InfoSec throughout hosts' lifetime may be excessively high or hard to predict	• ICNs may provide enhanced autonomy and memory, protection against social engineering, and other societal benefits
	Installers	• Equipment, training, and procedures to ensure InfoSec during implantation may be prohibitively costly	• Installation to enhance capacities of key individuals may have InfoSec benefits for society as a whole
	Hosts	• Loss of autonomy or personal identity • Loss of confidentiality, integrity, and availability of neural info processing	• Devices may provide external memory backup, automated threat detection and response, and other InfoSec services
Enable Ongoing Use	Operators	• Existing InfoSec ISMSes may be incompatible with ICNs, requiring costly development of new systems	• ICNs may allow better monitoring of device users and remote or automated provision of InfoSec services
	Maintainers	• Complexity of responding to ongoing evolution of threats and liability for failures may be too great	• Maintaining installed ICNs is needed for InfoSec, regardless of the aims and legality of the ICNs' installation

(Leftmost vertical label: ACTORS IN THE PROCESS OF DEVELOPING AND ADOPTING ICNS WHO...)

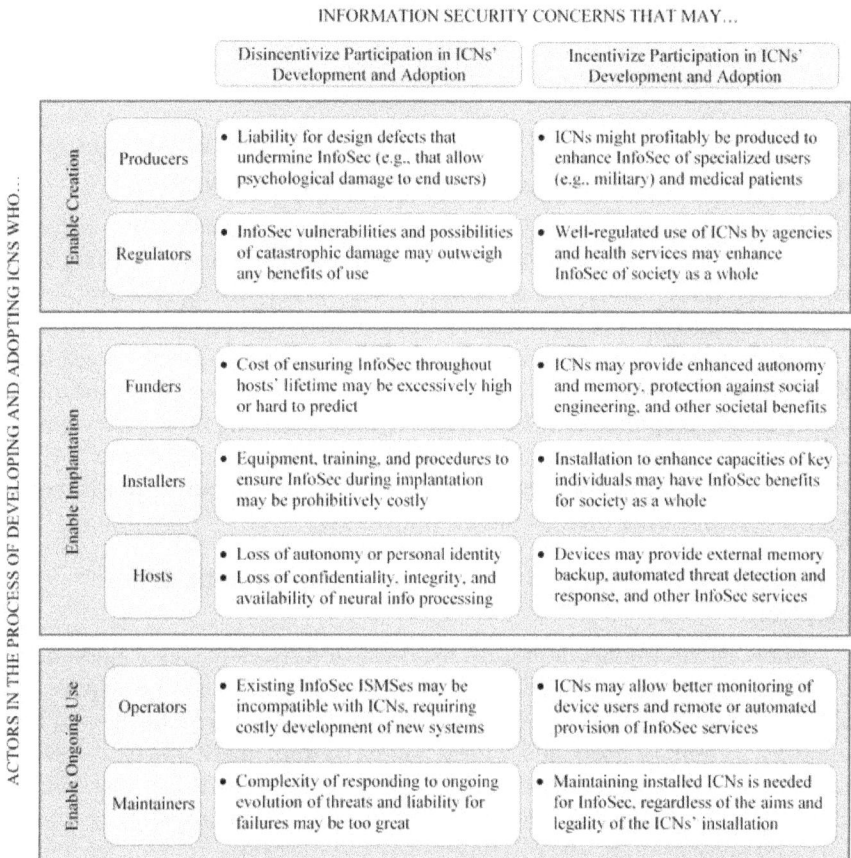

Figure 1. Examples of InfoSec-related concerns synthesized from InfoSec standards and literature that could potentially disincentivize or incentivize participation of seven key types of actors whose involvement is necessary in order for implantable cognitive neuroprostheses (ICNs) to be developed and deployed.

These resources do not focus specifically on the InfoSec questions that arise with the use of ICNs. However, those questions have been explored from an academic perspective in works such as those by McGee,[32] Denning et al.,[33] Koops and Leenes,[34] Kosta and Bowman,[35] and Gladden.[36] By interpreting the

---

[32] McGee (2008).

[33] Denning et al. (2009).

[34] Koops & Leenes (2012).

[35] Kosta & Bowman, "Implanting Implications: Data Protection Challenges Arising from the Use of Human ICT Implants" (2012).

[36] Gladden (2015).

published standards in light of such scholarship, it is possible to identify specific InfoSec concerns of relevance to the stakeholders whose participation is required for the implementation of ICNs.

Formulating a conceptual framework for InfoSec concerns as an impediment or impetus to the development of ICNs

In order to identify ways in which InfoSec concerns can either drive or impede the adoption of ICNs, we propose a conceptual framework that incorporates two dimensions: 1) the chain of actors who participate in the development and adoption of such technologies; and 2) their disincentivization or incentivization to participate in that process as a result of InfoSec considerations. Note that many other factors may influence whether actors decide to pursue the development of ICNs, including ethical, legal, public policy, financial, and operational considerations; the framework formulated here only attempts to identify those factors relating to information security. We can consider the framework's dimensions in more detail.

First dimension: actors in the process of neuroprosthetic devices' adoption

Review of the InfoSec literature for medical devices makes it possible to identify seven types of stakeholders whose participation will be required in order for any implantable neuroprosthetic technology to be developed and deployed in human hosts and whose failure to implement effective InfoSec measures could potentially result in injury or death for an ICN's host.[37] These actors include: 1) the designers and manufacturers of neuroprosthetic hardware and software (i.e., its 'producers'); 2) the government agencies and licensing bodies that must authorize the use of cognitive neuroprostheses in order for it to be legal (the technology's 'regulators'); 3) the government health services and private insurers that bear the cost of such devices' surgical implantation and ongoing maintenance ('funders'); 4) hospitals, clinics, and physicians who assess individual patients and perform the implantation of neuroprosthetic devices ('installers'); 5) the human subjects in whom neuroprosthetic devices are implanted but who may or may not actually operate the devices ('hosts'); 6) the typically institutional service providers that manage devices' connections to external systems and may remotely manage the devices themselves (their 'operators'); and 7) the providers of physical maintenance and upgrades, software updates, and additional functionality for neuroprosthetic devices already in use (their 'maintainers').

---

[37] *Content of Premarket Submissions* (2014), p. 3; Gladden (2015), pp. 109-10; *Postmarket Management of Cybersecurity* (2016), p. 10.

Collectively, the first two types of stakeholders (producers and regulators) can be understood as enabling the *creation* of implantable cognitive neuroprosthetic devices; the following three types (funders, installers, and hosts) as enabling their *implantation;* and the final two types (operators and maintainers) as enabling their *ongoing use.*

Second dimension: disincentivization or incentivization of participation in adoption process

For a given actor, InfoSec concerns may provide the actor with either disincentives or incentives (or both) to participate in the development and adoption of ICNs.

Discussion of potential InfoSec-related disincentives and incentives for each of the actor types to participate in ICN development

By combining both dimensions, a two-dimensional framework is created; Figure 1 presents such a framework that has been populated with sample InfoSec concerns drawn from industry standards and other literature. We can now explore conceptually how for each of the potential actors in the process, InfoSec concerns can create either a disincentive or incentive for the actor to participate in the development and adoption of ICNs.

Producers: designers and manufacturers of hardware and software

Designers and manufacturers are largely responsible for the InfoSec characteristics of ICNs.[38] The reliance of implantable neuroprostheses on mobile, wireless, and networked technologies places them at significant danger for the embedding of malicious code and other attacks that can exploit vulnerabilities in such technologies.[39] InfoSec breaches could have fatal consequences for the human hosts of ICNs;[40] large-scale catastrophic InfoSec failures attributable to a manufacturer could result in massive fines and remediation costs, irreparable reputation brand damage, and even bankruptcy.[41] Producers may thus decide that the risks inherent in producing ICNs outweigh any possible benefits.

Moreover, the unique nature of ICNs may create contradictory InfoSec-related design requirements which are infeasible for manufacturers to satisfy

[38] *Content of Premarket Submissions* (2014), p. 1.

[39] *ISO 27799* (2008), p. 47; *NIST SP 1800-1* (2016), Part a, p. 1; *Content of Premarket Submissions* (2014), p. 4.

[40] *ISO 27799* (2008), p. 47.

[41] "Security Risk Assessment Framework for Medical Devices" (2014), p. 16.

simultaneously. For example, devices allowing access to neural functions must be maximally secure while at the same time granting full and immediate access to medical personnel in case of an emergency.[42] Similarly, for ICNs that store data in a biological or biomimetic neural network[43] or which transmit data through synaptic connections with biological neurons, it may be impossible to utilize the InfoSec best practice of encrypting data[44] without destroying the information's availability and utility.

Despite these concerns, though, it is possible that some individuals or organizations may wish to employ ICNs precisely in order to enhance their own information security or to protect that of others. In such a case, InfoSec considerations would constitute a factor driving demand for ICNs, which could make their development and production profitable and desirable for device designers and manufacturers. Such potential uses for individuals include strengthening the agency of users whose autonomy has been reduced by disorders such as Parkinson's disease,[45] restoring the memory mechanisms of individuals suffering from Alzheimer's disease or other neurological disorders,[46] and providing the ability to record and 'play back' audiovisual experiences at will with perfect fidelity.[47] Potential uses for organizations include augmenting the brains of military personnel to aid in their work of gathering and processing intelligence and engaging in cyberwarfare and combat operations[48] and to enhance the availability of sensory information and memories by reducing their need for sleep.[49]

Regulators: agencies and licensing bodies authorizing device adoption

Regulatory agencies may be hesitant to approve the use of ICNs – especially for purposes of elective enhancement – if their InfoSec characteristics create a grave and widespread danger of psychological, physical, economic, or social harm for their users without counterbalancing benefits. However, regulators may be willing to authorize at least limited development of ICNs

---

[42] *Content of Premarket Submissions* (2014), p. 4.

[43] See Merkel et al. (2007); Rutten et al., "Neural Networks on Chemically Patterned Electrode Arrays: Towards a Cultured Probe" (2007); Stieglitz (2007); and Gladden (2015), p. 31.

[44] *NIST SP 1800-1* (2016), Part e, p. 5.

[45] Van den Berg (2012); Gladden (2015), pp. 97, 150-51.

[46] Ansari et al. (2007); Han et al. (2009); Ramirez et al. (2013); McGee (2008); Warwick (2014), p. 267.

[47] Merkel et al. (2007); Robinett, "The consequences of fully understanding the brain" (2002); McGee (2008), p. 217; Gladden (2015), pp. 156-57.

[48] Schermer, "The Mind and the Machine. On the Conceptual and Moral Implications of Brain-Machine Interaction" (2009); Brunner & Schalk (2009); Gladden (2015), p. 34.

[49] Kourany (2013); Gladden (2015), p. 151.

if they potentially create new and more effective tools for use by police personnel to analyze crime-related data and combat cybercrime, by military personnel to gather intelligence and conduct cyberwarfare, or by the personnel of private enterprises to detect and combat corporate espionage and cyberattacks.[50] Regulation may also be desirable in order to create and enforce national or international InfoSec standards that, for example, allow emergency access to ICNs by medical personnel.[51]

Funders: government health services and insurers subsidizing device use

The ongoing and unpredictable costs of protecting ICNs' human hosts from cyberattacks throughout the rest of their lives and of caring for those rendered psychologically, physically, or economically damaged as a result of such attacks may contribute to decisions by public health services and insurers that subsidizing the implantation and use of ICNs – especially those employed for elective enhancement – is not a sound investment.

On the other hand, institutions such as national governments and large corporations may be willing to fund the use of ICNs by their own personnel if the devices would be utilized to enhance the information security of those institutions or the constituencies they serve – such as when used by specialized military, police, health care, or corporate business intelligence and InfoSec personnel.[52] Moreover, expenditures enabling the successful widespread use of ICNs to treat disorders such as Alzheimer's disease[53] could be understood as enhancing the 'information security' of significant populations within society (e.g., by increasing the integrity, availability, and utility of memories and other information available to affected individuals and the autonomy of such human beings as host-device systems) and could potentially be justified by government health services on the grounds of improving public health and generating long-term savings on health care costs.

---

[50] Gladden (2015), p. 111.

[51] Cho & Lee, "Biometric Based Secure Communications without Pre-Deployed Key for Biosensor Implanted in Body Sensor Networks" (2012); Freudenthal et al., "Practical techniques for limiting disclosure of RF-equipped medical devices" (2007); Gladden (2015), p. 273.

[52] "Bridging the Bio-Electronic Divide" (2016); Szoldra, "The government's top scientists have a plan to make military cyborgs" (2016).

[53] Ansari et al. (2007).

Installers: hospitals and physicians who implant devices

Small clinics or hospitals with great expertise in performing surgical procedures may not possess equivalent expertise in information security,[54] making it impossible for them to ensure adequate information security during the preparatory, surgical, and recovery stages of an ICN implantation.

However, the implantation of ICNs by hospitals and physicians to treat disorders such as Alzheimer's and Parkinson's diseases,[55] treat emotional and psychological disorders,[56] and regulate levels of conscious alertness[57] could help fulfill their duty of care by enhancing the availability and integrity of patients' information and the autonomy of the patients' host-device systems. Possession of ICNs by a hospital's medical personnel could also enhance the availability of information for those personnel by, e.g., providing instantaneous, hands-free access to online reference texts[58] or real-time advice from other medical personnel.[59] It has been estimated that effective InfoSec practices in fields like health care can increase organizational performance by up to 2%;[60] if the use of ICNs by medical personnel would enhance their institutions' InfoSec performance, there may thus be managerial and financial incentives for their deployment, beyond any directly health-related rationales.

Hosts: human subjects and end users of neuroprosthetic devices

The human hosts of ICNs subject themselves to the potential introduction of computer viruses, worms, or malware[61] into their own cognitive processes and make their own thoughts and memories potential targets for attacks by hackers and other adversaries.[62] InfoSec failures relating to a host's ICN could result in a loss of autonomy and personal identity; psychological, physical, economic, or social harm; or potentially even the host's death.[63]

At the same time, particular human beings may have an incentive to acquire and utilize ICNs in order to combat the effects of Alzheimer's disease,

[54] *ISO 27799* (2008), p. v.
[55] Ansari et al. (2007); Van den Berg (2012).
[56] McGee (2008), p. 217.
[57] Claussen & Hofmann (2012); Kourany (2013), pp. 992-93.
[58] Gladden (2015), pp. 33, 156-57; McGee (2008).
[59] Rao et al. (2014); Gladden (2015), pp. 32-33.
[60] *ISO 27799* (2008), p. vi.
[61] *ISO 27799* (2008), p. 45; "Cybersecurity for Medical Devices and Hospital Networks: FDA Safety Communication" (2013), p. 1.
[62] *ISO 27799* (2008), p. 45; Denning et al. (2009); Gladden (2015).
[63] *ISO 27799* (2008), p. 5; Gladden (2015), pp. 145-68.

Parkinson's disease, emotional disorders, sleep disorders, and other conditions that negatively impact the integrity and availability of memories stored within their brains and the integrity and autonomy of the ongoing information-processing activities of the individuals' minds.[64] Individuals may also be able to use ICNs to enhance their information security beyond what is naturally possible for human beings – such as by artificially increasing the quantity and quality of external information accessible to their minds[65] or enhancing their 'internal' memory capacity beyond natural limits.[66]

Operators: managers of systems that monitor and control devices

ICNs may create residual risks that the operators of ICN systems are not able to mitigate through the implementation of compensating controls and which may endanger ICNs' 'essential clinical performance.'[67] For example, given an ICN's implantable nature, it may be impossible to maintain a secure physical perimeter around the device[68] and protect it from electromagnetic radiation and other potentially disruptive environmental emissions.[69] For ICNs that store and process data in the form of a biological or biomimetic neural network, it may be impractical or even impossible to regularly back up the devices' data in its entirety to a location that is physically secure in order to ensure its long-term availability.[70] Potential operators may also decide not to deploy or support ICNs due to the fact that the organizations' standard InfoSec practices cannot be applied to such devices. For example, operators of a health information system might typically limit network bandwidth for a compromised device or throttle its functionality in order to prevent it from degrading system services otherwise misusing system resources;[71] such a response may be impermissible if it would endanger the human host of a compromised ICN – who may not even be responsible for his or her device's excessive resource demand.

On the other hand, public health services may choose to operate ICN systems precisely in order to enhance the information security of patients suffering from cognitive disorders that disrupt the brain's ability to store or use

---

[64] Ansari et al. (2007); Van den Berg (2012); McGee (2008), p. 217; Claussen & Hofmann (2012); Kourany (2013), pp. 992-93; Soussou & Berger (2008); Gladden (2015), pp. 26-27.

[65] Koops & Leenes (2012); Merkel et al. (2007); Robinett (2002); McGee (2008), p. 217; Gladden (2015), pp. 156-57.

[66] Spohrer (2002); McGee (2008); Warwick (2014), p. 267; Gladden (2015), pp. 33, 148.

[67] See *Postmarket Management of Cybersecurity* (2016), pp. 9, 15.

[68] *ISO 27799* (2008), p. 29.

[69] *ISO 27799* (2008), p. 30.

[70] *ISO 27799* (2008), p. 32; Gladden (2015), p. 236.

[71] *ISO 27799* (2008), p. 46.

information.[72] Operators of ICN systems might also include government military or police agencies, large corporations, or other institutions for which maximizing information security and combatting InfoSec threats is a critical organizational objective; in particular, personnel augmented by such devices could be more effective at gathering and analyzing intelligence and protecting organizations from cyberattacks.[73] In the case of individual senior political figures or corporate executives, implantation of an ICN may be warranted in order to counteract the effects of Alzheimer's disease, Parkinson's disease, or other cognitive disorders that could impair the individuals' information security and thereby imperil the mission of the institutions in which they work.[74]

Maintainers: providers of software updates and physical maintenance services

Organizations (including third-party businesses) that provide physical maintenance services, antivirus software and updates, and other applications, upgrades, or accessories to expand the functionality of ICNs may be constrained in their ability to access necessary device functions and data due to legal restrictions regarding the privacy of personal health information[75] that bind the devices' installers and operators. Moreover, maintenance errors by third-party service providers can open a device to attacks[76] and create liability for those service providers. Such service providers recognize that a 'masquerade' committed by their own personnel to obtain unauthorized information relating to an ICN (either for financial reasons, to advance hacktivism, out of curiosity, or for other purposes) is also a very real danger;[77] in the case of ICNs, the chance that such InfoSec breaches would cause severe psychological or physical harm to a device's host creates risks that service providers may be unwilling to bear.

Regardless of how and why ICNs have been implanted, though, the provision of effective maintenance and upgrade services is necessary in order to protect their users' lives and ensure their information security[78] – thus creating a potentially profitable market for such services. In the absence of regular maintenance and upgrades, ICNs would be vulnerable to new and evolving threats – which is an especially critical problem in the case of devices that are so closely integrated with their hosts' brain functions.

---

[72] Ansari et al. (2007); Han et al. (2009); Ramirez et al. (2013); McGee (2008); Warwick (2014), p. 267; Soussou & Berger (2008).

[73] Schermer (2009); Brunner & Schalk (2009); Gladden (2015), p. 34.

[74] See Gladden (2015), pp. 144, 213, 216-17.

[75] *ISO 27799* (2008), p. 24.

[76] *ISO 27799* (2008), p. 48.

[77] *ISO 27799* (2008), p. 45.

[78] *Postmarket Management of Cybersecurity* (2016).

Conclusion

Many factors determine whether and how quickly particular new biotechnologies are developed and deployed. There is a widespread presumption that the need to ensure information security for organizations and individuals can create obstacles that *impede* or *disallow* the adoption of sensitive biotechnologies but that it cannot *accelerate* or *facilitate* the adoption of such technologies. It is rarely acknowledged by researchers, regulators, or industry practitioners that the desire for information security might itself potentially help drive the development and implementation of technologies such as implantable cognitive neuroprostheses. Thus the Medical Device Privacy Consortium argues, for example, that information security concerns "threaten to disrupt critical information flows to and from medical device companies,"[79] and the FDA contends that effective cybersecurity is needed to safeguard the functionality of implantable devices.[80] In the policy statements, standards, and outreach campaigns of such leading bodies there is no hint that the converse might also be true – i.e., that properly designed and functioning ICNs and other implantable devices might be deployed precisely for the purpose of safeguarding and enhancing information security of individual users, organizations, or sizeable populations within human society.

By applying the framework developed in this paper to analyze issues raised in industry standards and scholarly literature, we have shown that when 'information security' is appropriately understood in its full sense of assuring the confidentiality, integrity, availability, possession, authenticity, utility, distinguishability, rejectability, and autonomy of information and information systems, for each of the actors involved in the process of developing and deploying ICNs it is possible for InfoSec concerns to serve either as an obstacle that discourages an actor from taking part *or* as a driving factor that encourages an actor to participate in the development and adoption of ICNs. This is true despite – or perhaps because of – the fact that among all forms of implantable devices, ICNs are those that are most intimately integrated with the neural circuitry of their human hosts and which are able to most directly participate in cognitive processes that are critical for their hosts' psychological and physical well-being. It is our hope that conceptual frameworks such as the one developed here can serve as a basis for further theoretical and empirical studies to explore the ways in which InfoSec concerns can either hinder or impel the adoption of ICNs and other potentially revolutionary biotechnologies.

---

[79] "Welcome," Medical Device Privacy Consortium (2016).
[80] *Content of Premarket Submissions* (2014), p. 1.

References

Abrams, Jerold J. "Pragmatism, Artificial Intelligence, and Posthuman Bioethics: Shusterman, Rorty, Foucault." *Human Studies* 27, no. 3 (2004): 241-58.

Al-Hudhud, Ghada. "On Swarming Medical Nanorobots." *International Journal of Bio-Science & Bio-Technology* 4, no. 1 (2012): 75-90.

Ameen, Moshaddique Al, Jingwei Liu, and Kyungsup Kwak. "Security and Privacy Issues in Wireless Sensor Networks for Healthcare Applications." *Journal of Medical Systems* 36, no. 1 (2010): 93-101.

Ankarali, Z.E., Q.H. Abbasi, A.F. Demir, E. Serpedin, K. Qaraqe, and H. Arslan. "A Comparative Review on the Wireless Implantable Medical Devices Privacy and Security." In *2014 EAI 4th International Conference on Wireless Mobile Communication and Healthcare (Mobihealth)*, 246-49, 2014.

Ansari, Sohail, K. Chaudhri, and K. Al Moutaery. "Vagus Nerve Stimulation: Indications and Limitations." In *Operative Neuromodulation*, edited by Damianos E. Sakas and Brian A. Simpson, pp. 281-86. Acta Neurochirurgica Supplements 97/2. Springer Vienna, 2007.

Armando, Alessandro, Gabriele Costa, Alessio Merlo, and Luca Verderame. "Formal Modeling and Automatic Enforcement of Bring Your Own Device Policies." *International Journal of Information Security* (2014): 1-18.

Ayaz, Hasan, Patricia A. Shewokis, Scott Bunce, Maria Schultheis, and Banu Onaral. "Assessment of Cognitive Neural Correlates for a Functional Near Infrared-Based Brain Computer Interface System." In *Foundations of Augmented Cognition. Neuroergonomics and Operational Neuroscience*, edited by Dylan D. Schmorrow, Ivy V. Estabrooke, and Marc Grootjen, pp. 699-708. Lecture Notes in Computer Science 5638. Springer Berlin Heidelberg, 2009.

Baars, Bernard J. *In the Theater of Consciousness.* New York, NY: Oxford University Press, 1997.

Baddeley, Alan. "The episodic buffer: a new component of working memory?" *Trends in cognitive sciences* 4, no. 11 (2000): 417-23.

Badmington, Neil. "Cultural Studies and the Posthumanities," edited by Gary Hall and Claire Birchall. *New Cultural Studies: Adventures in Theory*, pp. 260-72. Edinburgh: Edinburgh University Press, 2006.

Baudrillard, Jean. *Simulacra and Simulation.* Ann Arbor: University of Michigan Press, 1994.

Bendle, Mervyn F. "Teleportation, cyborgs and the posthuman ideology." *Social Semiotics* 12, no. 1 (2002): 45-62.

Benedict, M., and H. Schlieter. "Governance Guidelines for Digital Healthcare Ecosystems," in *EHealth2015 – Health Informatics Meets EHealth: Innovative Health Perspectives: Personalized Health*, pp. 233-40. 2015.

Bergamasco, S., M. Bon, and P. Inchingolo. "Medical data protection with a new generation of hardware authentication tokens." In *IFMBE Proceedings MEDICON 2001*, edited by R. Magjarevic, S. Tonkovic, V. Bilas, and I. Lackovic, pp. 82-85. IFMBE, 2001.

Birbaumer, Niels, and Klaus Haagen. "Restoration of Movement and Thought from Neuroelectric and Metabolic Brain Activity: Brain-Computer Interfaces (BCIs)." In *Intelligent Computing Everywhere*, edited by Alfons J. Schuster, pp. 129-52. Springer London, 2007.

Birnbacher, Dieter. "Posthumanity, Transhumanism and Human Nature." In *Medical Enhancement and Posthumanity*, edited by Bert Gordijn and Ruth Chadwick, pp. 95-106. The International Library of Ethics, Law and Technology 2. Springer Netherlands, 2008.

Borkar, Shekhar. "Designing reliable systems from unreliable components: the challenges of transistor variability and degradation." *Micro, IEEE* 25, no. 6 (2005): 10-16.

Borton, D. A., Y.-K. Song, W. R. Patterson, C. W. Bull, S. Park, F. Laiwalla, J. P. Donoghue, and A. V. Nurmikko. "Implantable Wireless Cortical Recording Device for Primates." In *World Congress on Medical Physics and Biomedical Engineering, September 7-12, 2009, Munich, Germany*, edited by Olaf Dössel and Wolfgang C. Schlegel, pp. 384-87. IFMBE Proceedings 25/9. Springer Berlin Heidelberg, 2009.

Bostrom, Nick. "Why I Want to Be a Posthuman When I Grow Up." In *Medical Enhancement and Posthumanity*, edited by Bert Gordijn and Ruth Chadwick, pp. 107-36. The International Library of Ethics, Law and Technology 2. Springer Netherlands, 2008.

Bostrom, Nick, and Anders Sandberg. "Cognitive Enhancement: Methods, Ethics, Regulatory Challenges." *Science and Engineering Ethics* 15, no. 3 (2009): 311-41.

Bowman, Diana M., Mark N. Gasson, and Eleni Kosta. "The Societal Reality of That Which Was Once Science Fiction." In *Human ICT Implants: Technical, Legal and Ethical Considerations*, edited by Mark N. Gasson, Eleni Kosta, and Diana M. Bowman, pp. 175-79. Information Technology and Law Series 23. T. M. C. Asser Press, 2012.

Brey, Philip. "Ethical Aspects of Information Security and Privacy." In *Security, Privacy, and Trust in Modern Data Management*, edited by Milan Petković and Willem Jonker, pp. 21-36. Data-Centric Systems and Applications. Springer Berlin Heidelberg, 2007.

"Bridging the Bio-Electronic Divide." Defense Advanced Research Projects Agency, January 19, 2016. http://www.darpa.mil/news-events/2015-01-19. Accessed May 6, 2016.

Brunner, Peter, and Gerwin Schalk. "Brain-Computer Interaction." In *Foundations of Augmented Cognition. Neuroergonomics and Operational Neuroscience*, edited by Dylan D. Schmorrow, Ivy V. Estabrooke, and Marc Grootjen, pp. 719-23. Lecture Notes in Computer Science 5638. Springer Berlin Heidelberg, 2009.

Buller, Tom. "Neurotechnology, Invasiveness and the Extended Mind." *Neuroethics* 6, no. 3 (2011): 593-605.

Calverley, D.J. "Imagining a non-biological machine as a legal person." *AI & SOCIETY* 22, no. 4 (2008): 523-37.

Campbell, Courtney S., James F. Keenan, David R. Loy, Kathleen Matthews, Terry Winograd, and Laurie Zoloth. "The Machine in the Body: Ethical and Religious Issues in the Bodily Incorporation of Mechanical Devices." In *Altering Nature*, edited by B. Andrew Lustig, Baruch A. Brody, and Gerald P. McKenny, pp. 199-257. Philosophy and Medicine 98. Springer Netherlands, 2008.

Cervera-Paz, Francisco Javier, and M. J. Manrique. "Auditory Brainstem Implants: Past, Present and Future Prospects." In *Operative Neuromodulation*, edited by Damianos E. Sakas and Brian A. Simpson, pp. 437-42. Acta Neurochirurgica Supplements 97/2. Springer Vienna, 2007.

Chadwick, Ruth. "Therapy, Enhancement and Improvement." In *Medical Enhancement and Posthumanity*, edited by Bert Gordijn and Ruth Chadwick, pp. 25-37. The International Library of Ethics, Law and Technology 2. Springer Netherlands, 2008.

Chaudhry, Peggy E., Sohail S. Chaudhry, Ronald Reese, and Darryl S. Jones. "Enterprise Information Systems Security: A Conceptual Framework." In *Re-Conceptualizing Enterprise Information Systems*, edited by Charles Møller and Sohail Chaudhry, pp. 118-28. Lecture Notes in Business Information Processing 105. Springer Berlin Heidelberg, 2012.

Cho, Kwantae, and Dong Hoon Lee. "Biometric Based Secure Communications without Pre-Deployed Key for Biosensor Implanted in Body Sensor Networks." In *Information Security Applications*, edited by Souhwan Jung and Moti Yung, pp. 203-18. Lecture Notes in Computer Science 7115. Springer Berlin Heidelberg, 2012.

Church, George M., Yuan Gao, and Sriram Kosuri. "Next-generation digital information storage in DNA." *Science* 337, no. 6102 (2012): 1628.

Clark, S.S., and K. Fu. "Recent Results in Computer Security for Medical Devices." In *Wireless Mobile Communication and Healthcare*, edited by K.S. Nikita, J.C. Lin, D.I. Fotiadis, and M.-T. Arredondo Waldmeyer, pp. 111-18. Lecture Notes of the Institute for Computer Sciences, Social Informatics and Telecommunications Engineering 83. Springer Berlin Heidelberg, 2012.

Claussen, Jens Christian, and Ulrich G. Hofmann. "Sleep, Neuroengineering and Dynamics." *Cognitive Neurodynamics* 6, no. 3 (2012): 211-14.

Clowes, Robert W. "The Cognitive Integration of E-Memory." *Review of Philosophy and Psychology* 4, no. 1 (2013): 107-33.

Coeckelbergh, Mark. "From Killer Machines to Doctrines and Swarms, or Why Ethics of Military Robotics Is Not (Necessarily) About Robots." *Philosophy & Technology* 24, no. 3 (2011): 269-78.

Coles-Kemp, Lizzie, and Marianthi Theoharidou. "Insider Threat and Information Security Management." In *Insider Threats in Cyber Security*, edited by Christian W. Probst, Jeffrey Hunker, Dieter Gollmann, and Matt Bishop, pp. 45-71. Advances in Information Security 49. Springer US, 2010.

Content of Premarket Submissions for Management of Cybersecurity in Medical Devices: Guidance for Industry and Food and Drug Administration Staff. Silver Spring, MD: US Food and Drug Administration, 2014.

Cosgrove, G.R. "Session 6: Neuroscience, brain, and behavior V: Deep brain stimulation." Meeting of the President's Council on Bioethics. Washington, DC, June 24-25, 2004. https://bioethicsarchive.georgetown.edu/pcbe/transcripts/june04/session6.html. Accessed June 12, 2015.

"Cybersecurity for Medical Devices and Hospital Networks: FDA Safety Communication." U.S. Food and Drug Administration, June 13, 2013. http://www.fda.gov/MedicalDevices/Safety/AlertsandNotices/ucm356423.htm. Accessed May 3, 2016.

Dardick, Glenn. "Cyber Forensics Assurance." In *Proceedings of the 8th Australian Digital Forensics Conference*, pp. 57-64. Research Online, 2010.

Datteri, E. "Predicting the Long-Term Effects of Human-Robot Interaction: A Reflection on Responsibility in Medical Robotics." *Science and Engineering Ethics* 19, no. 1 (2013): 139-60.

Delac, Kresimir, and Mislav Grgic. "A Survey of Biometric Recognition Methods." In *Proceedings of the 46th International Symposium on Electronics in Marine, ELMAR 2004*, pp. 184-93. IEEE, 2004.

Denning, Tamara, Alan Borning, Batya Friedman, Brian T. Gill, Tadayoshi Kohno, and William H. Maisel. "Patients, pacemakers, and implantable defibrillators: Human values and security for wireless implantable medical devices." In *Proceedings of the SIGCHI Conference on Human Factors in Computing Systems*, pp. 917-26. ACM, 2010.

Denning, Tamara, Kevin Fu, and Tadayoshi Kohno. "Absence Makes the Heart Grow Fonder: New Directions for Implantable Medical Device Security." 3rd USENIX Workshop on Hot Topics in Security (HotSec 2008). San Jose, CA, July 29, 2008.

Denning, Tamara, Yoky Matsuoka, and Tadayoshi Kohno. "Neurosecurity: Security and Privacy for Neural Devices." *Neurosurgical Focus* 27, no. 1 (2009): E7.

Donchin, Emanuel, and Yael Arbel. "P300 Based Brain Computer Interfaces: A Progress Report." In *Foundations of Augmented Cognition. Neuroergonomics and Operational Neuroscience*, edited by Dylan D. Schmorrow, Ivy V. Estabrooke, and Marc Grootjen, pp. 724-31. Lecture Notes in Computer Science 5638. Springer Berlin Heidelberg, 2009.

Dormer, Kenneth J. "Implantable electronic otologic devices for hearing rehabilitation." In *Handbook of Neuroprosthetic Methods*, edited by Warren E. Finn and Peter G. LoPresti, pp. 237-60. Boca Raton: CRC Press, 2003.

Drongelen, Wim van, Hyong C. Lee, and Kurt E. Hecox. "Seizure Prediction in Epilepsy." In *Neural Engineering*, edited by Bin He, pp. 389-419. Bioelectric Engineering. Springer US, 2005.

Dudai, Yadin. "The Neurobiology of Consolidations, Or, How Stable Is the Engram?" *Annual Review of Psychology* 55 (2004): 51-86.

Durand, Dominique M., Warren M. Grill, and Robert Kirsch. "Electrical Stimulation of the Neuromuscular System." In *Neural Engineering*, edited by Bin He, pp. 157-91. Bioelectric Engineering. Springer US, 2005.

Dvorsky, George. "What may be the world's first cybernetic hate crime unfolds in French McDonald's." io9, July 17, 2012. http://io9.com/5926587/what-may-be-the-worlds-first-cybernetic-hate-crime-unfolds-in-french-mcdonalds. Accessed July 22, 2015.

Edlinger, Günter, Cristiano Rizzo, and Christoph Guger. "Brain Computer Interface." In *Springer Handbook of Medical Technology*, edited by Rüdiger Kramme, Klaus-Peter Hoffmann, and Robert S. Pozos, pp. 1003-17. Springer Berlin Heidelberg, 2011.

Erler, Alexandre. "Does Memory Modification Threaten Our Authenticity?" *Neuroethics* 4, no. 3 (2011): 235-49.

Evans, Dave. "The Internet of Everything: How More Relevant and Valuable Connections Will Change the World." Cisco Internet Solutions Business Group: Point of View, 2012. https://www.cisco.com/web/about/ac79/docs/innov/IoE.pdf. Accessed December 16, 2015.

Fairclough, S.H. "Physiological Computing: Interfacing with the Human Nervous System." In *Sensing Emotions*, edited by J. Westerink, M. Krans, and M. Ouwerkerk, pp. 1-20. Philips Research Book Series 12. Springer Netherlands, 2010.

Fernandes, Diogo A. B., Liliana F. B. Soares, João V. Gomes, Mário M. Freire, and Pedro R. M. Inácio. "Security Issues in Cloud Environments: A Survey." *International Journal of Information Security* 13, no. 2 (2013): 113-70.

Ferrando, Francesca. "Posthumanism, Transhumanism, Antihumanism, Metahumanism, and New Materialisms: Differences and Relations." *Existenz: An International Journal in Philosophy, Religion, Politics, and the Arts* 8, no. 2 (Fall 2013): 26-32.

FIPS PUB 199: Standards for Security Categorization of Federal Information and Information Systems. Gaithersburg, MD: National Institute of Standards and Technology, 2004.

Fleischmann, Kenneth R. "Sociotechnical Interaction and Cyborg–Cyborg Interaction: Transforming the Scale and Convergence of HCI." *The Information Society* 25, no. 4 (2009): 227-35.

Fountas, Kostas N., and J. R. Smith. "A Novel Closed-Loop Stimulation System in the Control of Focal, Medically Refractory Epilepsy." In *Operative Neuromodulation*, edited by Damianos E. Sakas and Brian A. Simpson, pp. 357-62. Acta Neurochirurgica Supplements 97/2. Springer Vienna, 2007.

Freudenthal, Eric, Ryan Spring, and Leonardo Estevez. "Practical techniques for limiting disclosure of RF-equipped medical devices." In *Engineering in Medicine and Biology Workshop, 2007 IEEE Dallas*, pp. 82-85. IEEE, 2007.

Friedenberg, Jay. *Artificial Psychology: The Quest for What It Means to Be Human*. Philadelphia: Psychology Press, 2008.

Fukuyama, Francis. *Our Posthuman Future: Consequences of the Biotechnology Revolution*. New York: Farrar, Straus, and Giroux, 2002.

Gärtner, Armin. "Communicating Medical Systems and Networks." In *Springer Handbook of Medical Technology*, edited by Rüdiger Kramme, Klaus-Peter Hoffmann, and Robert S. Pozos, pp. 1085-93. Springer Berlin Heidelberg, 2011.

Gasson, M.N., Kosta, E., and Bowman, D.M. "Human ICT Implants: From Invasive to Pervasive." In *Human ICT Implants: Technical, Legal and Ethical Considerations*, edited by Mark N. Gasson, Eleni Kosta, and Diana M. Bowman, pp. 1-8. Information Technology and Law Series 23. T. M. C. Asser Press, 2012.

Gasson, M.N. "Human ICT Implants: From Restorative Application to Human Enhancement." In *Human ICT Implants: Technical, Legal and Ethical Considerations*, edited by Mark N. Gasson, Eleni Kosta, and Diana M. Bowman, pp. 11-28. Information Technology and Law Series 23. T. M. C. Asser Press, 2012.

Gasson, M.N. "ICT Implants." In *The Future of Identity in the Information Society*, edited by S. Fischer-Hübner, P. Duquenoy, A. Zuccato, and L. Martucci, pp. 287-95. Springer US, 2008.

Gerhardt, Greg A., and Patrick A. Tresco. "Sensor Technology." In *Brain-Computer Interfaces*, pp. 7-29. Springer Netherlands, 2008.

Gladden, Matthew E. "Cryptocurrency with a Conscience: Using Artificial Intelligence to Develop Money that Advances Human Ethical Values." *Annales: Ethics in Economic Life* vol. 18, no. 4 (2015): 85-98.

Gladden, Matthew E. "Cybershells, Shapeshifting, and Neuroprosthetics: Video Games as Tools for Posthuman 'Body Schema (Re)Engineering'." Keynote presentation at the Ogólnopolska Konferencja Naukowa Dyskursy Gier Wideo, Facta Ficta / AGH, Kraków, June 6, 2015.

Gladden, Matthew E. "The Diffuse Intelligent Other: An Ontology of Nonlocalizable Robots as Moral and Legal Actors." In *Social Robots: Boundaries, Potential, Challenges*, edited by Marco Nørskov, pp. 177-98. Farnham: Ashgate, 2016.

Gladden, Matthew E. "Enterprise Architecture for Neurocybernetically Augmented Organizational Systems: The Impact of Posthuman Neuroprosthetics on the Creation of Strategic, Structural, Functional, Technological, and Sociocultural Alignment." Thesis project, MBA in Innovation and Data Analysis. Warsaw: Institute of Computer Science, Polish Academy of Sciences, 2016.

Gladden, Matthew E. "A Fractal Measure for Comparing the Work Effort of Human and Artificial Agents Performing Management Functions." In *Position Papers of the 2014 Federated Conference on Computer Science and Information Systems*, edited by Maria Ganzha, Leszek Maciaszek, Marcin Paprzycki, pp. 219-26. Annals of Computer Science and Information Systems 3. Polskie Towarzystwo Informatyczne, 2014.

Gladden, Matthew E. *The Handbook of Information Security for Advanced Neuroprosthetics*. Indianapolis: Synthypnion Academic, 2015.

Gladden, Matthew E. "Information Security Concerns as a Catalyst for the Development of Implantable Cognitive Neuroprostheses." In *9th Annual EuroMed Academy of Business (EMAB) Conference: Innovation, Entrepreneurship and Digital Ecosystems (EUROMED 2016) Book of Proceedings*, edited by Demetris Vrontis, Yaakov Weber, and Evangelos Tsoukatos, pp. 891-904. Engomi: EuroMed Press, 2016.

Gladden, Matthew E. "Managing the Ethical Dimensions of Brain-Computer Interfaces in eHealth: An SDLC-based Approach." In *9th Annual EuroMed Academy of Business (EMAB) Conference: Innovation, Entrepreneurship and Digital Ecosystems (EUROMED 2016) Book of Proceedings*, edited by Demetris Vrontis, Yaakov Weber, and Evangelos Tsoukatos, pp. 876-90. Engomi: EuroMed Press, 2016.

Gladden, Matthew E. "Neural Implants as Gateways to Digital-Physical Ecosystems and Posthuman Socioeconomic Interaction." In *Digital Ecosystems: Society in the Digital Age*, edited by Łukasz Jonak, Natalia Juchniewicz, and Renata Włoch, pp. 85-98. Warsaw: Digital Economy Lab, University of Warsaw, 2016.

Gladden, Matthew E. *Neuroprosthetic Supersystems Architecture*. Indianapolis: Synthypnion Academic, 2017.

Gladden, Matthew E. *Sapient Circuits and Digitalized Flesh: The Organization as Locus of Technological Posthumanization*. Indianapolis: Defragmenter Media, 2016.

Gladden, Matthew E. "Utopias and Dystopias as Cybernetic Information Systems: Envisioning the Posthuman Neuropolity." *Creatio Fantastica* nr 3 (50) (2015).

Graham, Elaine. *Representations of the Post/Human: Monsters, Aliens and Others in Popular Culture*. Manchester: Manchester University Press, 2002.

Greenberg, Andy. "Cyborg Discrimination? Scientist Says McDonald's Staff Tried To Pull Off His Google-Glass-Like Eyepiece, Then Threw Him Out." Forbes, July 17, 2012. http://www.forbes.com/sites/andygreenberg/2012/07/17/cyborg-discrimination-scientist-says-mcdonalds-staff-tried-to-pull-off-his-google-glass-like-eyepiece-then-threw-him-out/. Accessed July 22, 2015.

Grodzinsky, F.S., K.W. Miller, and M.J. Wolf. "Developing Artificial Agents Worthy of Trust: 'Would You Buy a Used Car from This Artificial Agent?'" *Ethics and Information Technology* 13, no. 1 (2011): 17-27.

Grottke, M., H. Sun, R.M. Fricks, and K.S. Trivedi. "Ten fallacies of availability and reliability analysis." In *Service Availability*, pp. 187-206. Lecture Notes in Computer Science 5017. Springer Berlin Heidelberg, 2008.

Guidance for Industry - Cybersecurity for Networked Medical Devices Containing Off-the-Shelf (OTS) Software. Silver Spring, MD: US Food and Drug Administration, 2005.

Gunkel, David J. *The Machine Question: Critical Perspectives on AI, Robots, and Ethics*. Cambridge, MA: The MIT Press, 2012.

Gunther, N. J. "Time—the zeroth performance metric." In *Analyzing Computer System Performance with Perl::PDQ*, 3-46. Berlin: Springer, 2005.

Halperin, Daniel, Tadayoshi Kohno, Thomas S. Heydt-Benjamin, Kevin Fu, and William H. Maisel. "Security and privacy for implantable medical devices." *Pervasive Computing, IEEE* 7, no. 1 (2008): 30-39.

Han, J.-H., S.A. Kushner, A.P. Yiu, H.-W. Hsiang, T. Buch, A. Waisman, B. Bontempi, R.L. Neve, P.W. Frankland, and S.A. Josselyn. "Selective Erasure of a Fear Memory." *Science* 323, no. 5920 (2009): 1492-96.

Hansen, Jeremy A., and Nicole M. Hansen. "A Taxonomy of Vulnerabilities in Implantable Medical Devices." In *Proceedings of the Second Annual Workshop on Security and Privacy in Medical and Home-Care Systems*, pp. 13-20. ACM, 2010.

Hanson, R. "If uploads come first: The crack of a future dawn." *Extropy* 6, no. 2 (1994): 10-15.

Haraway, Donna. "A Manifesto for Cyborgs: Science, Technology, and Socialist Feminism in the 1980s." *Socialist Review* 15, no. 2 (1985): 65-107.

Haraway, Donna. *Simians, Cyborgs, and Women: The Reinvention of Nature*. New York: Routledge, 1991.

Harrison, Ian. "IEC80001 and Future Ramifications for Health Systems Not Currently Classed as Medical Devices." In *Making Systems Safer*, edited by Chris Dale and Tom Anderson, pp. 149-71. Springer London, 2010.

Hatfield, B., A. Haufler, and J. Contreras-Vidal. "Brain Processes and Neurofeedback for Performance Enhancement of Precision Motor Behavior." In *Foundations of Augmented Cognition. Neuroergonomics and Operational Neuroscience*, edited by Dylan D. Schmorrow, Ivy V. Estabrooke, and Marc Grootjen, pp. 810-17. Lecture Notes in Computer Science 5638. Springer Berlin Heidelberg, 2009.

Hayles, N. Katherine. *How We Became Posthuman: Virtual Bodies in Cybernetics, Literature, and Informatics*. Chicago: University of Chicago Press, 1999.

Heersmink, Richard. "Embodied Tools, Cognitive Tools and Brain-Computer Interfaces." *Neuroethics* 6, no. 1 (2011): 207-19.

Hei, Xiali, and Xiaojiang Du. "Biometric-based two-level secure access control for implantable medical devices during emergencies." In *INFOCOM, 2011 Proceedings IEEE*, pp. 346-350. IEEE, 2011.

Hellström, T. "On the Moral Responsibility of Military Robots." *Ethics and Information Technology* 15, no. 2 (2013): 99-107.

Herbrechter, Stefan. *Posthumanism: A Critical Analysis*. London: Bloomsbury, 2013. [Kindle edition.]

Hern, Alex. "Hacker fakes German minister's fingerprints using photos of her hands." The Guardian, December 30, 2014. http://www.theguardian.com/technology/2014/dec/30/hacker-fakes-german-ministers-fingerprints-using-photos-of-her-hands. Accessed July 24, 2015.

Heylighen, Francis. "The Global Brain as a New Utopia." In *Renaissance der Utopie. Zukunftsfiguren des 21. Jahrhunderts*, edited by R. Maresch and F. Rötzer. Frankfurt: Suhrkamp, 2002.

Hildebrandt, Mireille, and Bernhard Anrig. "Ethical Implications of ICT Implants." In *Human ICT Implants: Technical, Legal and Ethical Considerations*, edited by Mark N. Gasson, Eleni Kosta, and Diana M. Bowman, pp. 135-58. Information Technology and Law Series 23. T. M. C. Asser Press, 2012.

Hochmair, Ingeborg. "Cochlear Implants: Facts." MED-EL, September 2013. http://www.medel.com/cochlear-implants-facts. Accessed December 8, 2016.

Hoffmann, Klaus-Peter, and Silvestro Micera. "Introduction to Neuroprosthetics." In *Springer Handbook of Medical Technology*, edited by Rüdiger Kramme, Klaus-Peter Hoffmann, and Robert S. Pozos, pp. 785-800. Springer Berlin Heidelberg, 2011.

Humphreys, L., J. M. Ferrández, and E. Fernández. "Long Term Modulation and Control of Neuronal Firing in Excitable Tissue Using Optogenetics." In *Foundations on Natural and Artificial Computation*, edited by José Manuel Ferrández, José Ramón Álvarez Sánchez, Félix de la Paz, and F. Javier Toledo, pp. 266-73. Lecture Notes in Computer Science 6686. Springer Berlin Heidelberg, 2011.

IEC 80001: Application of risk management for IT-networks incorporating medical devices, Parts 1 through 2-7. ISO/TC 215. Geneva: IEC, 2010-15.

Illes, Judy. *Neuroethics: Defining the Issues in Theory, Practice, and Policy.* Oxford University Press, 2006.

ISO 27799:2008, Health informatics – Information security management in health using ISO/IEC 27002. ISO/TC 215. Geneva: ISO/IEC, 2008.

ISO/IEC 27001:2013, Information technology – Security techniques – Information security management systems – Requirements. ISO/IEC JTC 1/SC 27. Geneva: ISO/IEC, 2013.

ISO/IEC 27002:2013, Information technology – Security techniques – Code of practice for information security controls. ISO/IEC JTC 1/SC 27. Geneva: ISO/IEC, 2013.

ISO/TR 11633-1:2009, Health informatics – Information security management for remote maintenance of medical devices and medical information systems – Part 1: Requirements and risk analysis. ISO/TC 215. Geneva: ISO, 2009.

ISO/TR 11633-2:2009, Health informatics – Information security management for remote maintenance of medical devices and medical information systems – Part 2: Implementation of an information security management system (ISMS). ISO/TC 215. Geneva: ISO, 2009.

Josselyn, Sheena A. "Continuing the Search for the Engram: Examining the Mechanism of Fear Memories." *Journal of Psychiatry & Neuroscience : JPN* 35, no. 4 (2010): 221-28.

Kelly, Kevin. "A Taxonomy of Minds." *The Technium,* February 15, 2007. http://kk.org/thetechnium/a-taxonomy-of-m/. Accessed January 25, 2016.

Kelly, Kevin. "The Landscape of Possible Intelligences." *The Technium,* September 10, 2008. http://kk.org/thetechnium/the-landscape-o/. Accessed January 25, 2016.

Kelly, Kevin. *Out of Control: The New Biology of Machines, Social Systems and the Economic World.* Basic Books, 1994.

Kirkpatrick, K. "Legal Issues with Robots." *Communications of the ACM* 56, no. 11 (2013): 17-19.

KleinOsowski, A., Ethan H. Cannon, Phil Oldiges, and Larry Wissel. "Circuit design and modeling for soft errors." *IBM Journal of Research and Development* 52, no. 3 (2008): 255-63.

Kłoda-Staniecko, Bartosz. "Ja, Cyborg. Trzy porządki, jeden byt. Podmiot jako fuzja biologii, kultury i technologii" ("I, Cyborg. Three Orders, One Being. Subject as a Fusion of Nature, Culture and Technology"). In *Człowiek w relacji do zwierząt, roślin i maszyn w kulturze: Tom I: Aspekt posthumanistyczny i transhumanistyczny,* edited by Justyny Tymienieckiej-Suchanek. Uniwersytet Śląski, 2015.

Koch, K. P. "Neural Prostheses and Biomedical Microsystems in Neurological Rehabilitation." In *Operative Neuromodulation,* edited by Damianos E. Sakas, Brian A. Simpson, and Elliot S. Krames, pp. 427-34. Acta Neurochirurgica Supplements 97/1. Springer Vienna, 2007.

Koebler, Jason. "FCC Cracks Down on Cell Phone 'Jammers': The FCC says illegal devices that block cell phone signals could pose security risk." *U.S. News & World Report,* October 17, 2012. http://www.usnews.com/news/articles/2012/10/17/fcc-cracks-down-on-cell-phone-jammers. Accessed July 22, 2015.

Koene, Randal A. "Embracing Competitive Balance: The Case for Substrate-Independent Minds and Whole Brain Emulation." In *Singularity Hypotheses,* edited by Amnon H. Eden, James H. Moor, Johnny H. Søraker, and Eric Steinhart, pp. 241-67. The Frontiers Collection. Springer Berlin Heidelberg, 2012.

Koops, B.-J., and R. Leenes. "Cheating with Implants: Implications of the Hidden Information Advantage of Bionic Ears and Eyes." In *Human ICT Implants: Technical, Legal and Ethical*

Considerations, edited by Mark N. Gasson, Eleni Kosta, and Diana M. Bowman, pp. 113-34. Information Technology and Law Series 23. T. M. C. Asser Press, 2012.

Kosta, E., and D.M. Bowman, "Implanting Implications: Data Protection Challenges Arising from the Use of Human ICT Implants." In *Human ICT Implants: Technical, Legal and Ethical Considerations*, edited by Mark N. Gasson, Eleni Kosta, and Diana M. Bowman, pp. 97-112. Information Technology and Law Series 23. T. M. C. Asser Press, 2012.

Kourany, J.A. "Human Enhancement: Making the Debate More Productive." *Erkenntnis* 79, no. 5 (2013): 981-98.

Kowalewska, Agata. "Symbionts and Parasites – Digital Ecosystems." In *Digital Ecosystems: Society in the Digital Age*, edited by Łukasz Jonak, Natalia Juchniewicz, and Renata Włoch, pp. 73-84. Warsaw: Digital Economy Lab, University of Warsaw, 2016.

Kraemer, Felicitas. "Me, Myself and My Brain Implant: Deep Brain Stimulation Raises Questions of Personal Authenticity and Alienation." *Neuroethics* 6, no. 3 (2011): 483-97. doi:10.1007/s12152-011-9115-7.

Kuflik, A. "Computers in Control: Rational Transfer of Authority or Irresponsible Abdication of Autonomy?" *Ethics and Information Technology* 1, no. 3 (1999): 173-84.

Lebedev, M. "Brain-Machine Interfaces: An Overview." *Translational Neuroscience* 5, no. 1 (2014): 99-110.

Leder, Felix, Tillmann Werner, and Peter Martini. "Proactive Botnet Countermeasures: An Offensive Approach." In *The Virtual Battlefield: Perspectives on Cyber Warfare*, volume 3, edited by Christian Czosseck and Kenneth Geers, pp. 211-25. IOS Press, 2009.

Lee, Giljae, Andréa Matsunaga, Salvador Dura-Bernal, Wenjie Zhang, William W. Lytton, Joseph T. Francis, and José AB Fortes. "Towards Real-Time Communication between in Vivo Neurophysiological Data Sources and Simulator-Based Brain Biomimetic Models." *Journal of Computational Surgery* 3, no. 1 (2014): 1-23.

Li, S., F. Hu, and G. Li, "Advances and Challenges in Body Area Network." In *Applied Informatics and Communication*, edited by J. Zhan, pp. 58-65. Communications in Computer and Information Science 22. Springer Berlin Heidelberg, 2011.

Lind, Jürgen. "Issues in agent-oriented software engineering." In *Agent-Oriented Software Engineering*, pp. 45-58. Springer Berlin Heidelberg, 2001.

Linsenmeier, Robert A. "Retinal Bioengineering." In *Neural Engineering*, edited by Bin He, pp. 421-84. Bioelectric Engineering. Springer US, 2005.

Longuet-Higgins, H.C. "Holographic Model of Temporal Recall." *Nature* 217, no. 5123 (1968): 104.

Lucivero, Federica, and Guglielmo Tamburrini. "Ethical Monitoring of Brain-Machine Interfaces." *AI & SOCIETY* 22, no. 3 (2007): 449-60.

Ma, Ting, Ying-Ying Gu, and Yuan-Ting Zhang. "Circuit Models for Neural Information Processing." In *Neural Engineering*, edited by Bin He, pp. 333-65. Bioelectric Engineering. Springer US, 2005.

MacVittie, Kevin, Jan Halámek, Lenka Halámková, Mark Southcott, William D. Jemison, Robert Lobel, and Evgeny Katz. "From 'cyborg' lobsters to a pacemaker powered by implantable biofuel cells." *Energy & Environmental Science* 6, no. 1 (2013): 81-86.

Maguire, Gerald Q., and Ellen M. McGee. "Implantable brain chips? Time for debate." *Hastings Center Report* 29, no. 1 (1999): 7-13.

Maj, Krzysztof. "Rational Technotopia vs. Corporational Dystopia in 'Deus Ex: Human Revolution' Gameworld." His Master's Voice: Utopias and Dystopias in Audiovisual Culture. Facta Ficta Research Centre / Jagiellonian University, Kraków, March 24, 2015.

Mak, Stephen. "Ethical Values for E-Society: Information, Security and Privacy." In *Ethics and Policy of Biometrics*, edited by Ajay Kumar and David Zhang, pp. 96-101. Lecture Notes in Computer Science 6005. Springer Berlin Heidelberg, 2010.

Masani, Kei, and Milos R. Popovic. "Functional Electrical Stimulation in Rehabilitation and Neurorehabilitation." In *Springer Handbook of Medical Technology*, edited by Rüdiger Kramme, Klaus-Peter Hoffmann, and Robert S. Pozos, pp. 877-96. Springer Berlin Heidelberg, 2011.

McCormick, Michael. "Data Theft: A Prototypical Insider Threat." In *Insider Attack and Cyber Security*, edited by Salvatore J. Stolfo, Steven M. Bellovin, Angelos D. Keromytis, Shlomo Hershkop, Sean W. Smith, and Sara Sinclair, pp. 53-68. Advances in Information Security 39. Springer US, 2008.

McCullagh, P., G. Lightbody, J. Zygierewicz, and W.G. Kernohan. "Ethical Challenges Associated with the Development and Deployment of Brain Computer Interface Technology." *Neuroethics* 7, no. 2 (2013): 109-22.

McGee, E.M. "Bioelectronics and Implanted Devices." In *Medical Enhancement and Posthumanity*, edited by Bert Gordijn and Ruth Chadwick, pp. 207-24. The International Library of Ethics, Law and Technology 2. Springer Netherlands, 2008.

McGrath, Michael J., and Cliodhna Ní Scanaill. "Regulations and Standards: Considerations for Sensor Technologies." In *Sensor Technologies*, pp. 115-35. Apress, 2013.

McIntosh, Daniel. "The Transhuman Security Dilemma." *Journal of Evolution and Technology* 21, no. 2 (2010): 32-48.

Medical Enhancement and Posthumanity, edited by Bert Gordijn and Ruth Chadwick. The International Library of Ethics, Law and Technology 2. Springer Netherlands, 2008.

Meloy, Stuart. "Neurally Augmented Sexual Function." In *Operative Neuromodulation*, edited by Damianos E. Sakas, Brian A. Simpson, and Elliot S. Krames, pp. 359-63. Acta Neurochirurgica Supplements 97/1. Springer Vienna, 2007.

Merkel, R., G. Boer, J. Fegert, T. Galert, D. Hartmann, B. Nuttin, and S. Rosahl. "Central Neural Prostheses." In *Intervening in the Brain: Changing Psyche and Society*, pp. 117-60. Ethics of Science and Technology Assessment 29. Springer Berlin Heidelberg, 2007.

Miah, Andy. "A Critical History of Posthumanism." In *Medical Enhancement and Posthumanity*, edited by Bert Gordijn and Ruth Chadwick, pp. 71-94. The International Library of Ethics, Law and Technology 2. Springer Netherlands, 2008.

Miller, Kai J., and Jeffrey G. Ojemann. "A Simple, Spectral-Change Based, Electrocorticographic Brain–Computer Interface." In *Brain-Computer Interfaces*, edited by Bernhard Graimann, Gert Pfurtscheller, and Brendan Allison, pp. 241-58. The Frontiers Collection. Springer Berlin Heidelberg, 2009.

Miller, Jr., Gerald Alva. "Conclusion: Beyond the Human: Ontogenesis, Technology, and the Posthuman in Kubrick and Clarke's 2001." In *Exploring the Limits of the Human through Science Fiction*, pp. 163-90. American Literature Readings in the 21st Century. Palgrave Macmillan US, 2012.

Mitcheson, Paul D. "Energy harvesting for human wearable and implantable bio-sensors." In *Engineering in Medicine and Biology Society (EMBC), 2010 Annual International Conference of the IEEE*, pp. 3432-36. IEEE, 2010.

Mizraji, Eduardo, Andrés Pomi, and Juan C. Valle-Lisboa. "Dynamic Searching in the Brain." *Cognitive Neurodynamics* 3, no. 4 (2009): 401-14.

Moravec, Hans. *Mind Children: The Future of Robot and Human Intelligence*. Cambridge: Harvard University Press, 1990.

Moxon, Karen A. "Neurorobotics." In *Neural Engineering*, edited by Bin He, pp. 123-55. Bioelectric Engineering. Springer US, 2005.

Negoescu, R. "Conscience and Consciousness in Biomedical Engineering Science and Practice." In *International Conference on Advancements of Medicine and Health Care through Technology*, edited by Simona Vlad, Radu V. Ciupa, and Anca I. Nicu, pp. 209-14. IFMBE Proceedings 26. Springer Berlin Heidelberg, 2009.

NIST Special Publication 800-33: Underlying Technical Models for Information Technology Security. Edited by Gary Stoneburner. Gaithersburg, Maryland: National Institute of Standards & Technology, 2001.

NIST Special Publication 800-37, Revision 1: Guide for Applying the Risk Management Framework to Federal Information Systems: A Security Life Cycle Approach. Joint Task Force Transformation Initiative. Gaithersburg, Maryland: National Institute of Standards & Technology, 2010.

NIST Special Publication 800-53, Revision 4: Security and Privacy Controls for Federal Information Systems and Organizations. Joint Task Force Transformation Initiative. Gaithersburg, Maryland: National Institute of Standards & Technology, 2013.

NIST Special Publication 800-100: Information Security Handbook: A Guide for Managers. Edited by P. Bowen, J. Hash, and M. Wilson. Gaithersburg, Maryland: National Institute of Standards & Technology, 2006.

NIST Special Publication 1800-1: Securing Electronic Health Records on Mobile Devices (Draft), Parts a, b, c, d, and e. Edited by G. O'Brien, N. Lesser, B. Pleasant, S. Wang, K. Zheng, C. Bowers, K. Kamke, and L. Kauffman. Gaithersburg, Maryland: National Institute of Standards & Technology, 2015.

Ochsner, Beate, Markus Spöhrer, and Robert Stock. "Human, non-human, and beyond: cochlear implants in socio-technological environments." *NanoEthics* 9, no. 3 (2015): 237-50.

Overman, Stephenie. "Jamming Employee Phones Illegal." Society for Human Resource Management, May 9, 2014. http://www.shrm.org/hrdisciplines/technology/articles/pages/cell-phone-jamming.aspx. Accessed July 22, 2015.

Pająk, Robert. Email correspondence with the author, May 3, 2015.

Panoulas, Konstantinos J., Leontios J. Hadjileontiadis, and Stavros M. Panas. "Brain-Computer Interface (BCI): Types, Processing Perspectives and Applications." In *Multimedia Services in Intelligent Environments*, edited by George A. Tsihrintzis and Lakhmi C. Jain, pp. 299-321. Smart Innovation, Systems and Technologies 3. Springer Berlin Heidelberg, 2010.

Park, M.C., M.A. Goldman, T.W. Belknap, and G.M. Friehs. "The Future of Neural Interface Technology." In *Textbook of Stereotactic and Functional Neurosurgery*, edited by A.M. Lozano, P.L. Gildenberg, and R.R. Tasker, pp. 3185-3200. Heidelberg/Berlin: Springer, 2009.

Parker, Donn "Our Excessively Simplistic Information Security Model and How to Fix It." *ISSA Journal* (July 2010): 12-21.

Parker, Donn B. "Toward a New Framework for Information Security." In *The Computer Security Handbook*, fourth edition, edited by Seymour Bosworth and M. E. Kabay. John Wiley & Sons, 2002.

Passeraub, Ph A., and N. V. Thakor. "Interfacing Neural Tissue with Microsystems." In *Neural Engineering*, edited by Bin He, 49-83. Bioelectric Engineering. Springer US, 2005.

Patil, P.G., and D.A. Turner. "The Development of Brain-Machine Interface Neuroprosthetic Devices." *Neurotherapeutics* 5, no. 1 (2008): 137-46.

Pearce, David. "The Biointelligence Explosion." In *Singularity Hypotheses*, edited by A.H. Eden, J.H. Moor, J.H. Søraker, and E. Steinhart, pp. 199-238. The Frontiers Collection. Berlin/Heidelberg: Springer, 2012.

Polikov, Vadim S., Patrick A. Tresco, and William M. Reichert. "Response of brain tissue to chronically implanted neural electrodes." *Journal of Neuroscience Methods* 148, no. 1 (2005): 1-18.

Posthuman Bodies, edited by Judith Halberstam and Ira Livingstone. Bloomington, IN: Indiana University Press, 1995.

Postmarket Management of Cybersecurity in Medical Devices: Draft Guidance for Industry and Food and Drug Administration Staff. Silver Spring, MD: US Food and Drug Administration, 2016.

Pribram, K.H., and S.D. Meade. "Conscious Awareness: Processing in the Synaptodendritic Web – The Correlation of Neuron Density with Brain Size." *New Ideas in Psychology* 17, no. 3 (1999): 205-14.

Pribram, K.H. "Prolegomenon for a Holonomic Brain Theory." In *Synergetics of Cognition*, edited by Hermann Haken and Michael Stadler, pp. 150-84. Springer Series in Synergetics 45. Springer Berlin Heidelberg, 1990.

Principe, José C., and Dennis J. McFarland. "BMI/BCI Modeling and Signal Processing." In *Brain-Computer Interfaces*, pp. 47-64. Springer Netherlands, 2008.

Proudfoot, Diane. "Software Immortals: Science or Faith?" In *Singularity Hypotheses*, edited by Amnon H. Eden, James H. Moor, Johnny H. Søraker, and Eric Steinhart, pp. 367-92. The Frontiers Collection. Springer Berlin Heidelberg, 2012.

Qureshi, Mohmad Kashif. "Liveness detection of biometric traits." *International Journal of Information Technology and Knowledge Management* 4 (2011): 293-95.

Rahimi, Ali, Ben Recht, Jason Taylor, and Noah Vawter. "On the effectiveness of aluminium foil helmets: An empirical study." MIT, February 17, 2005. http://web.archive.org/web/20100708230258/http://people.csail.mit.edu/rahimi/helmet/. Accessed July 26, 2015.

Ramirez, S., X. Liu, P.-A. Lin, J. Suh, M. Pignatelli, R.L. Redondo, T.J. Ryan, and S. Tonegawa. "Creating a False Memory in the Hippocampus." *Science* 341, no. 6144 (2013): 387-91.

Rao, Umesh Hodeghatta, and Umesha Nayak. *The InfoSec Handbook*. New York: Apress, 2014.

Rao, R.P.N., A. Stocco, M. Bryan, D. Sarma, T.M. Youngquist, J. Wu, and C.S. Prat. "A direct brain-to-brain interface in humans." *PLoS ONE* 9, no. 11 (2014).

Rasmussen, Kasper Bonne, Claude Castelluccia, Thomas S. Heydt-Benjamin, and Srdjan Capkun. "Proximity-based access control for implantable medical devices." In *Proceedings of the 16th ACM conference on Computer and communications security*, pp. 410-19. ACM, 2009.

Robinett, W. "The consequences of fully understanding the brain." In *Converging Technologies for Improving Human Performance: Nanotechnology, Biotechnology, Information Technology and Cognitive Science*, edited by M.C. Roco and W.S. Bainbridge, pp. 166-70. National Science Foundation, 2002.

Roden, David. *Posthuman Life: Philosophy at the Edge of the Human*. Abingdon: Routledge, 2014.

Roosendaal, Arnold. "Carrying Implants and Carrying Risks; Human ICT Implants and Liability." In *Human ICT Implants: Technical, Legal and Ethical Considerations*, edited by Mark N.

Gasson, Eleni Kosta, and Diana M. Bowman, pp. 69-79. Information Technology and Law Series 23. T. M. C. Asser Press, 2012.

Roosendaal, Arnold. "Implants and Human Rights, in Particular Bodily Integrity." In *Human ICT Implants: Technical, Legal and Ethical Considerations*, edited by Mark N. Gasson, Eleni Kosta, and Diana M. Bowman, pp. 81-96. Information Technology and Law Series 23. T. M. C. Asser Press, 2012.

Rossebeø, J. E. Y., M. S. Lund, K. E. Husa, and A. Refsdal, "A conceptual model for service availability." In *Quality of Protection*, pp. 107-18. Advances in Information Security 23. Springer US, 2006.

Rotter, Pawel, Barbara Daskala, and Ramon Compañó. "Passive Human ICT Implants: Risks and Possible Solutions." In *Human ICT Implants: Technical, Legal and Ethical Considerations*, edited by Mark N. Gasson, Eleni Kosta, and Diana M. Bowman, pp. 55-62. Information Technology and Law Series 23. T. M. C. Asser Press, 2012.

Rotter, Pawel, and Mark N. Gasson. "Implantable Medical Devices: Privacy and Security Concerns." In *Human ICT Implants: Technical, Legal and Ethical Considerations*, edited by Mark N. Gasson, Eleni Kosta, and Diana M. Bowman, pp. 63-66. Information Technology and Law Series 23. T. M. C. Asser Press, 2012.

Rotter, Pawel, Barbara Daskala, Ramon Compañó, Bernhard Anrig, and Claude Fuhrer. "Potential Application Areas for RFID Implants." In *Human ICT Implants: Technical, Legal and Ethical Considerations*, edited by Mark N. Gasson, Eleni Kosta, and Diana M. Bowman, pp. 29-39. Information Technology and Law Series 23. T. M. C. Asser Press, 2012.

Rowlands, Mark. *Can Animals Be Moral?* Oxford: Oxford University Press, 2012.

Rubin, Charles T. "What Is the Good of Transhumanism?" In *Medical Enhancement and Posthumanity*, edited by Bert Gordijn and Ruth Chadwick, pp. 137-56. The International Library of Ethics, Law and Technology 2. Springer Netherlands, 2008.

Rutherford, Andrew, Gerasimos Markopoulos, Davide Bruno, and Mirjam Brady-Van den Bos. "Long-Term Memory: Encoding to Retrieval." In *Cognitive Psychology*, second edition, edited by Nick Braisby and Angus Gellatly, pp. 229-65. Oxford: Oxford University Press, 2012.

Rutten, W. L. C., T. G. Ruardij, E. Marani, and B. H. Roelofsen. "Neural Networks on Chemically Patterned Electrode Arrays: Towards a Cultured Probe." In *Operative Neuromodulation*, edited by Damianos E. Sakas and Brian A. Simpson, pp. 547-54. Acta Neurochirurgica Supplements 97/2. Springer Vienna, 2007.

Sakas, Damianos E., I. G. Panourias, and B. A. Simpson. "An Introduction to Neural Networks Surgery, a Field of Neuromodulation Which Is Based on Advances in Neural Networks Science and Digitised Brain Imaging." In *Operative Neuromodulation*, edited by Damianos E. Sakas and Brian A. Simpson, pp. 3-13. Acta Neurochirurgica Supplements 97/2. Springer Vienna, 2007.

Sandberg, Anders. "Ethics of brain emulations." *Journal of Experimental & Theoretical Artificial Intelligence* 26, no. 3 (2014): 439-57.

Sasse, Martina Angela, Sacha Brostoff, and Dirk Weirich. "Transforming the 'weakest link'—a human/computer interaction approach to usable and effective security." *BT technology journal* 19, no. 3 (2001): 122-31.

Schechter, Stuart. "Security that is Meant to be Skin Deep: Using Ultraviolet Micropigmentation to Store Emergency-Access Keys for Implantable Medical Devices." Microsoft Research, August 10, 2010. http://research.microsoft.com:8082/apps/pubs/default.aspx?id=135291. Accessed July 26, 2015.

Schermer, Maartje. "The Mind and the Machine. On the Conceptual and Moral Implications of Brain-Machine Interaction." *NanoEthics* 3, no. 3 (2009): 217-30.

"Security Risk Assessment Framework for Medical Devices." Washington, DC: Medical Device Privacy Consortium, 2014.

Shoniregun, Charles A., Kudakwashe Dube, and Fredrick Mtenzi. "Introduction to E-Healthcare Information Security." In *Electronic Healthcare Information Security*, pp. 1-27. Advances in Information Security 53. Springer US, 2010.

Soussou, Walid V., and Theodore W. Berger. "Cognitive and Emotional Neuroprostheses." In *Brain-Computer Interfaces*, pp. 109-23. Springer Netherlands, 2008.

Spohrer, Jim. "NBICS (Nano-Bio-Info-Cogno-Socio) Convergence to Improve Human Performance: Opportunities and Challenges." In *Converging Technologies for Improving Human Performance: Nanotechnology, Biotechnology, Information Technology and Cognitive Science*, edited by M.C. Roco and W.S. Bainbridge, pp. 101-17. Arlington, Virginia: National Science Foundation, 2002.

Srinivasan, G. R. "Modeling the cosmic-ray-induced soft-error rate in integrated circuits: an overview." *IBM Journal of Research and Development* 40, no. 1 (1996): 77-89.

Stahl, B. C. "Responsible Computers? A Case for Ascribing Quasi-Responsibility to Computers Independent of Personhood or Agency." *Ethics and Information Technology* 8, no. 4 (2006): 205-13.

Stieglitz, Thomas. "Restoration of Neurological Functions by Neuroprosthetic Technologies: Future Prospects and Trends towards Micro-, Nano-, and Biohybrid Systems." In *Operative Neuromodulation*, edited by Damianos E. Sakas, Brian A. Simpson, and Elliot S. Krames, pp. 435-42. Acta Neurochirurgica Supplements 97/1. Springer Vienna, 2007.

Szoldra, P. "The government's top scientists have a plan to make military cyborgs." Tech Insider, January 22, 2016. http://www.techinsider.io/darpa-neural-interface-2016-1. Accessed May 6, 2016.

Tadeusiewicz, Ryszard, Pawel Rotter, and Mark N. Gasson. "Restoring Function: Application Exemplars of Medical ICT Implants." In *Human ICT Implants: Technical, Legal and Ethical Considerations*, edited by Mark N. Gasson, Eleni Kosta, and Diana M. Bowman, pp. 41-51. Information Technology and Law Series 23. T. M. C. Asser Press, 2012.

Taira, Takaomi, and T. Hori. "Diaphragm Pacing with a Spinal Cord Stimulator: Current State and Future Directions." In *Operative Neuromodulation*, edited by Damianos E. Sakas, Brian A. Simpson, and Elliot S. Krames, pp. 289-92. Acta Neurochirurgica Supplements 97/1. Springer Vienna, 2007.

Tamburrini, Guglielmo. "Brain to Computer Communication: Ethical Perspectives on Interaction Models." *Neuroethics* 2, no. 3 (2009): 137-49.

Taylor, Dawn M. "Functional Electrical Stimulation and Rehabilitation Applications of BCIs." In *Brain-Computer Interfaces*, pp. 81-94. Springer Netherlands, 2008.

Thanos, Solon, P. Heiduschka, and T. Stupp. "Implantable Visual Prostheses." In *Operative Neuromodulation*, edited by Damianos E. Sakas and Brian A. Simpson, pp. 465-72. Acta Neurochirurgica Supplements 97/2. Springer Vienna, 2007.

Thonnard, Olivier, Leyla Bilge, Gavin O'Gorman, Seán Kiernan, and Martin Lee. "Industrial Espionage and Targeted Attacks: Understanding the Characteristics of an Escalating Threat." In *Research in Attacks, Intrusions, and Defenses*, edited by Davide Balzarotti, Salvatore J. Stolfo, and Marco Cova, pp. 64-85. Lecture Notes in Computer Science 7462. Springer Berlin Heidelberg, 2012.

Thorpe, Julie, Paul C. van Oorschot, and Anil Somayaji. "Pass-thoughts: authenticating with our minds." In *Proceedings of the 2005 Workshop on New Security Paradigms*, pp. 45-56. ACM, 2005.

Troyk, Philip R., and Stuart F. Cogan. "Sensory Neural Prostheses." In *Neural Engineering*, edited by Bin He, pp. 1-48. Bioelectric Engineering. Springer US, 2005.

Ullah, Sana, Henry Higgin, M. Arif Siddiqui, and Kyung Sup Kwak. "A Study of Implanted and Wearable Body Sensor Networks." In *Agent and Multi-Agent Systems: Technologies and Applications*, edited by Ngoc Thanh Nguyen, Geun Sik Jo, Robert J. Howlett, and Lakhmi C. Jain, pp. 464-73. Lecture Notes in Computer Science 4953. Springer Berlin Heidelberg, 2008.

U.S. Code, Title 44 (Public Printing and Documents), Subchapter III (Information Security), Section 3542 (Definitions), cited in *NIST Special Publication 800-37, Revision 1*.

Van den Berg, Bibi. "Pieces of Me: On Identity and Information and Communications Technology Implants." In *Human ICT Implants: Technical, Legal and Ethical Considerations*, edited by Mark N. Gasson, Eleni Kosta, and Diana M. Bowman, pp. 159-73. Information Technology and Law Series 23. T. M. C. Asser Press, 2012.

Vildjiounaite, Elena, Satu-Marja Mäkelä, Mikko Lindholm, Reima Riihimäki, Vesa Kyllönen, Jani Mäntyjärvi, and Heikki Ailisto. "Unobtrusive Multimodal Biometrics for Ensuring Privacy and Information Security with Personal Devices." In *Pervasive Computing*, edited by Kenneth P. Fishkin, Bernt Schiele, Paddy Nixon, and Aaron Quigley, pp. 187-201. Lecture Notes in Computer Science 3968. Springer Berlin Heidelberg, 2006.

Viola, M. V., and Aristides A. Patrinos. "A Neuroprosthesis for Restoring Sight." In *Operative Neuromodulation*, edited by Damianos E. Sakas and Brian A. Simpson, pp. 481-86. Acta Neurochirurgica Supplements 97/2. Springer Vienna, 2007.

Wager, K.A., F. Wickham Lee, and J.P. Glaser. *Health Care Information Systems: A Practical Approach for Health Care Management.* John Wiley & Sons, 2013.

Wallach, Wendell, and Colin Allen. *Moral machines: Teaching robots right from wrong.* Oxford University Press, 2008.

Warwick, K. "The Cyborg Revolution." *Nanoethics* 8 (2014): 263-73.

Weber, R. H., and R. Weber. "General Approaches for a Legal Framework." In *Internet of Things*, pp. 23-40. Springer Berlin/Heidelberg, 2010.

Weiland, James D., Wentai Liu, and Mark S. Humayun. "Retinal Prosthesis." *Annual Review of Biomedical Engineering* 7, no. 1 (2005): 361-401.

Weinberger, Sharon. "Mind Games." *Washington Post*, January 14, 2007. http://www.washingtonpost.com/wp-dyn/content/article/2007/01/10/AR2007011001399.html. Accessed July 26, 2015.

"Welcome." Medical Device Privacy Consortium. http://deviceprivacy.org. Accessed May 6, 2016.

Werkhoven, Peter. "Experience Machines: Capturing and Retrieving Personal Content." In *E-Content*, edited by Peter A. Bruck, Zeger Karssen, Andrea Buchholz, and Ansgar Zerfass, pp. 183-202. Springer Berlin Heidelberg, 2005.

Westlake, Philip R. "The possibilities of neural holographic processes within the brain." *Biological Cybernetics* 7, no. 4 (1970): 129-53.

Widge, A.S., C.T. Moritz, and Y. Matsuoka. "Direct Neural Control of Anatomically Correct Robotic Hands." In *Brain-Computer Interfaces*, edited by D.S. Tan and A. Nijholt, pp. 105-19. Human-Computer Interaction Series. London: Springer, 2010.

Wiener, Norbert. *Cybernetics: Or Control and Communication in the Animal and the Machine*, second edition. Cambridge, MA: The MIT Press, 1961. [Quid Pro ebook edition for Kindle, 2015.]

Wilkinson, Jeff, and Scott Hareland. "A cautionary tale of soft errors induced by SRAM packaging materials." *IEEE Transactions on Device and Materials Reliability* 5, no. 3 (2005): 428-33.

Wooldridge, M., and N. R. Jennings. "Intelligent agents: Theory and practice." *The Knowledge Engineering Review*, 10(2) (1995): 115-52.

Yampolskiy, Roman V. "The Universe of Minds." arXiv preprint, *arXiv:1410.0369 [cs.AI]*, October 1, 2014. http://arxiv.org/abs/1410.0369. Accessed January 25, 2016.

Yonck, Richard. "Toward a standard metric of machine intelligence." *World Future Review* 4, no. 2 (2012): 61-70.

Zamanian, Ali, and Cy Hardiman. "Electromagnetic radiation and human health: A review of sources and effects." *High Frequency Electronics* 4, no. 3 (2005): 16-26.

Zaród, Marcin. "Constructing Hackers. Professional Biographies of Polish Hackers." Digital Ecosystems. Digital Economy Lab, University of Warsaw, Warsaw, June 29, 2015.

Zebda, Abdelkader, S. Cosnier, J.-P. Alcaraz, M. Holzinger, A. Le Goff, C. Gondran, F. Boucher, F. Giroud, K. Gorgy, H. Lamraoui, and P. Cinquin. "Single glucose biofuel cells implanted in rats power electronic devices." *Scientific Reports* 3, article 1516 (2013).

Zhao, QiBin, LiQing Zhang, and Andrzej Cichocki. "EEG-Based Asynchronous BCI Control of a Car in 3D Virtual Reality Environments." *Chinese Science Bulletin* 54, no. 1 (2009): 78-87.

Zheng, Guanglou, Gengfa Fang, Mehmet Orgun, and Rajan Shankaran. "A Non-key based security scheme supporting emergency treatment of wireless implants." In *2014 IEEE International Conference on Communications (ICC)*, pp. 647-52. IEEE, 2014.

Zheng, Guanglou, Gengfa Fang, Mehmet Orgun, Rajan Shankaran, and Eryk Dutkiewicz. "Securing wireless medical implants using an ECG-based secret data sharing scheme." In *2014 14th International Symposium on Communications and Information Technologies (ISCIT)*, pp. 373-77. IEEE, 2014.

Zheng, Guanglou, Gengfa Fang, Rajan Shankaran, Mehmet Orgun, and Eryk Dutkiewicz. "An ECG-based secret data sharing scheme supporting emergency treatment of Implantable Medical Devices." In *2014 International Symposium on Wireless Personal Multimedia Communications (WPMC)*, pp. 624-28. IEEE, 2014.

Index

About the Author

Matthew E. Gladden is a management consultant and researcher whose work focuses on the organizational implications of emerging technologies such as those relating to artificial intelligence, social robotics, virtual reality, neuroprosthetic enhancement, and artificial life. He lectures internationally on the relationship of such posthumanizing technologies to organizational life, and his research has been published in journals such as the *International Journal of Contemporary Management, Annals of Computer Science and Information Systems, Informatyka Ekonomiczna / Business Informatics, Creatio Fantastica*, and *Annales: Ethics in Economic Life*, as well as by IOS Press, Ashgate Publishing, the Digital Economy Lab of the University of Warsaw, and the MIT Press. His books include *Sapient Circuits and Digitalized Flesh: The Organization as Locus of Technological Posthumanization* (2016); *Posthuman Management: Creating Effective Organizations in an Age of Social Robotics, Ubiquitous AI, Human Augmentation, and Virtual Worlds* (second edition, 2016); and *Neuroprosthetic Supersystems Architecture* (2017).

He is the founder and CEO of consulting firms NeuraXenetica LLC and Cognitive Firewall LLC. He previously served as Administrator of the Department of Psychology at Georgetown University and Associate Director of the Woodstock Theological Center and has also taught philosophical ethics and worked in computer game design. He is a member of ISACA, ISSA, and the Academy of Management and its divisions for Managerial and Organizational Cognition, Technology and Innovation Management, and Business Policy and Strategy. He completed his MBA in Innovation and Data Analysis at the Institute of Computer Science of the Polish Academy of Sciences and holds certificates in Advanced Business Management and Nonprofit Management from Georgetown University and a BA in Philosophy from Wabash College.

www.ingramcontent.com/pod-product-compliance
Lightning Source LLC
Chambersburg PA
CBHW022054210326
41519CB00054B/335